Longlining

Also of interest

Fish Catching Methods of the World
Third edition
A. Von Brandt
0 85238 125 5

Multilingual Dictionary of Fish and Fish Products
Fourth edition
OECD
0 85238 216 2

Fisheries Mismanagement
The Case of the North Atlantic Cod
R. Hannesson
0 85238 243 X

The Icelandic Fisheries
Evolution and Management of a Fishing Industry
R. Arnason
0 85238 210 3

Purse Seining Manual
M. Ben-Yami
0 85238 193 X

Fishermen's Handbook
P. Branson
0 85238 143 3

Fisheries Management in Crisis
K. Crean and D. Symes
0 85238 231 6

Longlining

Åsmund Bjordal and Svein Løkkeborg

Fishing News Books

Fishing News Books
A division of Blackwell Science Ltd
Editorial Offices:
Osney Mead, Oxford OX2 0EL
25 John Street, London WC1N 2BL
23 Ainslie Place, Edinburgh EH3 6AJ
238 Main Street, Cambridge
Massachusetts 02142, USA
54 University Street, Carlton
Victoria 3053, Australia

Other Editorial Offices:
Arnette Blackwell SA
224, Boulevard Saint Germain
75007 Paris, France

Blackwell Wissenschafts-Verlag GmbH
Kurfürstendamm 57
10707 Berlin, Germany

Zehetnergasse 6
A-1140 Wien
Austria

First published 1996

DISTRIBUTORS

Marston Book Services Ltd
PO Box 269
Abingdon
Oxon OX14 4YN
(*Orders:* Tel: 01865 206206
Fax: 01865 721205
Telex: 83355 MEDBOK G)

USA
Blackwell Science, Inc.
238 Main Street
Cambridge, MA 02142
(*Orders:* Tel: 800 215-1000
617 876-7000
Fax: 617 492-5263)

Canada
Copp Clark, Ltd
2775 Matheson Blvd East
Mississauga, Ontario
Canada, L4W 4P7
(*Orders:* Tel: 800 263-4374
905 238-6074)

Australia
Blackwell Science Pty Ltd
54 University Street
Carlton, Victoria 3053
(*Orders:* Tel: 03 9347-0300
Fax: 03 9349-3016)

A catalogue record for this title is available from the British Library

ISBN: 978-0-85238-200-4

Library of Congress
Cataloging-in-Publication Data
Bjordal, Åsmund.
Longlining/Åsmund Bjordal and Svein Løkkeborg.
p. cm.
Includes bibliographical references (p.) and index.

1. Longlining (Fisheries) I. Løkkeborg, Svein. II. Title.
SH344.6.L64B45 1996 95-47347
639.2′2–dc20 CIP

Contents

Preface

Despite the fact that longline fishing is one of our most traditional and widespread fish capture methods, there has been no comprehensive literature describing the principles and application of such fishing gear. The intention of this book is therefore to give the reader an introduction to different aspects of longlining, such as the history of the gear, the basic capture principles with related fish behaviour aspects, important longline target species, technical descriptions of the gear, vessels and equipment, and different modes of operating longlines, as well as to compare longline fishing with other fishing methods.

The book is aimed at a wide readership, ranging from professional longline fishermen to fisheries managers and students, as well as those among the general public with an interest in the subject.

In writing the book we soon realised that it would be difficult to make it comprehensive by describing the great variety of longline fishing on a global scale. We have therefore focused on the general principles of longlining. Our examples come mainly from the North Atlantic longline fisheries and stem from 25 years of longline research and development work at the Fish Capture Division of the Institute of Marine Research, Norway. For those with an interest in longlining beyond the coverage of this book, we hope that the list of selected literature provides a useful guide to further reading.

We are thankful to several people who have helped us by providing figures, photographs and statistics on longline fisheries from different parts of the world. In particular we would like to thank Stein Mortensen for providing the line drawings, Jofrid Øvredal for preparing diagrams, Anne Brit Skar Tysseland for preparing the tables and Tom Wray for useful comments on the manuscript. Finally, we would like to thank Blackwell Science for their extraordinary patience in the face of delays during the preparation of this book.

Å.B.
S.L.
Bergen,
January 1996

The Authors

Both authors are fisheries biologists presently working at the Institute of Marine Research in Bergen, Norway, Åsmund Bjordal as research director at the Department of Marine Resources, and Svein Løkkeborg as senior scientist at the Fish Capture Division.

Bjordal took his masters degree in marine biology at the University of California, Santa Barbara, and his Cand. real degree (1979, fisheries biology) at the University of Bergen, specialising in fish behaviour studies in relation to fishing gears, with a dissertation on the behaviour of Norway lobster towards baited pots. Except for a year at the Alaska Fisheries Science Center in Seattle, his academic career has been in Norway with the Institute of Fishery Technology Research and the Marine Research Institute in Bergen.

Løkkeborg received his academic education at the University of Bergen and received his Dr scient. degree in 1990, with a dissertation based on studies of fish behaviour in relation to natural and artificial longline baits. Apart from a year's stay at the Hatfield Marine Science Center in Newport, Oregon, his academic career has been with the Institute of Fishery Technology Research and the Marine Research Institute in Bergen.

Chapter 1
Introduction

Longlining is one of our major fish capture techniques. It is defined as a passive fishing method, i.e. the gear is stationary and the encounter between gear and fish is a result of fish movement towards the gear. As with pots, creels and traps, longlines are based on fish attraction by means of bait, which serves as the source of smell and taste stimuli to lure the fish to the gear and stimulate it to ingest the baited hook. The success of capture is then left to the ability of the hook to catch and retain the fish until it is brought on board the fishing vessel.

Longline gear is used all over the world, from small-scale artisanal fishing to modern mechanised longline operations. The longline is basically a very simple fishing gear, but there are great variations in gear construction, mode of operation and fishing strategy. Our intention is to give the reader a basic understanding of longline gear and longline fishing. Although we have tried to include examples of longline fishing from different parts of the world, most of the background material is taken from the north east Atlantic longline fisheries.

1.1 Basic definitions

As indicated by the name of the gear, the main part of it is a long (main) line to which shorter (branch) lines with baited hooks are attached. It is appropriate here to introduce the basic longline unit and definitions as these terms will be used frequently throughout the book.

The basic longline gear unit consists of four parts (see Fig. 1.1):

- The *mainline* (also called *groundline*) is the 'backbone' of the longline gear to which the snoods and hooks are attached at intervals
- The *snood* (also called *branchline*, *gangion*, or *gangline*)
- The *hook*
- The *bait*

The term *hook spacing* should also be introduced at this point. This is a general term for the distance between neighbouring hooks, measured as the distance between the attachment points of two neighbouring snoods on the stretched mainline.

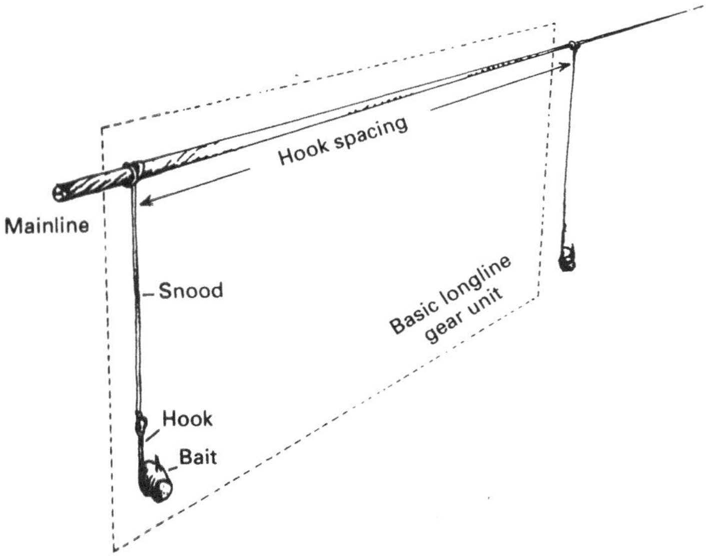

Fig. 1.1 The basic unit of longline gear: mainline, snood, hook and bait. (Drawing by Stein Mortensen.)

This basic longline unit is, however, a rather theoretical one; the unit in practical rigging and use of longlines is the so-called *skate*, which is a specified length of mainline which varies according to the type of fishing, with a certain number of snoods and hooks (100–500) spaced at regular intervals. Several skates can be linked to make a *fleet* of longline gear. A fleet may vary in length from a few hundred metres in small-scale coastal fishing to more than 50 km in large-scale oceanic longline fisheries.

1.2 Longline fishing from a historical perspective

Any development of fishing gear is based on a knowledge of fish behaviour. Historically we do not know when man started to utilise bait for fishing, but this most probably originated from observations of fish being attracted to discarded offal. We do know that ancient man used hooks for fishing as fish hooks are commonly found during archaeological excavations, indicating that hook and line fishing dates back to the Stone Age. Various materials have been used for making hooks throughout history: stone, bone, horn, sea shells, wood and metal. Until man learned the art of working with metal, the properties of the material restricted the shape of the hooks. Examples of different typical ancient hook designs are shown in Fig. 1.2.

It is interesting to note that fish-hook shapes developed differently in the Atlantic and Pacific areas. Figure 1.2 shows ancient hooks of

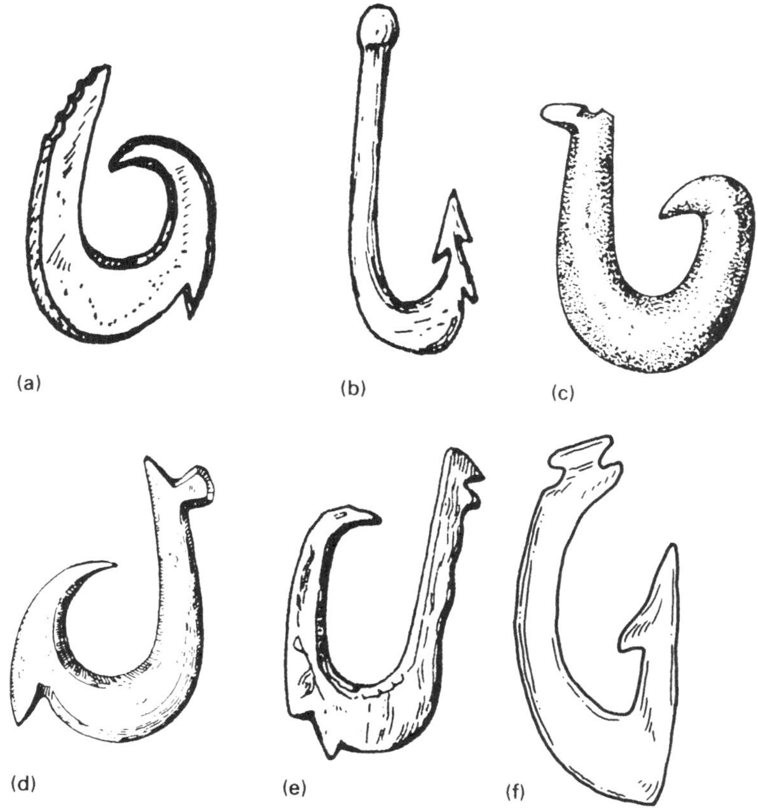

Fig. 1.2 Ancient hook designs: (a) shell hook, Santa Barbara, California; (b) bone hook, Vinda, Yugoslavia; (c) stone hook, Pitcairn; (d) whale tooth hook, Hawaii; (e) reindeer hook, Japan; (f) hook from Agder, Norway, considered to be 4000 years old. (Redrawn after Hurum, 1977.)

Atlantic origin which are mostly variations of the so-called J-hook. In the Pacific, and particularly the oceanic islands, the basic shape was the circle. This is possibly a consequence of the main material being used, namely sea shells which naturally lend themselves to this form. These hooks are also called rotating hooks, and were traditionally barbless. The fish were caught through the wedging of the jaw bone between the curved point and shank of the hook, rather than by penetration and securing of the hook by the barb as is the principle with J-hooks (see Fig. 1.3). However, the traditional circle-shaped hook with a barb has become very popular in commercial longlining in recent years, as has the EZ-baiter hook, which is a hybrid design incorporating the benefits of both the traditional circle and J-hook shapes (see Figs. 5.6 to 5.8).

The hook is of little use without a line, and the development of lines must therefore have accompanied the ancient hook designs. However,

Fig. 1.3 Hooking principle of the rotating (circle-shaped) hook. (Redrawn after Johannes, 1981.)

as lines were made from degradable materials, they are found only rarely during excavations. Hooks and lines were used by man for fish capture several thousand years BC as indicated by the age of the hooks shown in Fig. 1.2, some of which are considered to be more than 4000 years old. Hook and line fishing is also mentioned in Norse mythology in the story about the god Thor trying to catch the great sea serpent with hook and line using an ox head as bait. (The serpent, with a total length that encircled the globe, struck the baited hook but after a terrible fight Thor had to abandon landing it. So the great serpent is still there waiting to be caught – possibly with more modern hook and line gear.) Early depictions of hook and line fishing can be found among Scandinavian rock carvings dating from the Bronze Age. Figure 1.4 shows ancient fishermen using handlines.

Fig. 1.4 Bronze Age rock carving depicting hook and line fishing (Bohuslän, Swedish west coast, about 1000 BC). (Drawing by Å. Fredsjö, from Hurum, 1977.)

Hook and line fishing with one or a few hooks on a handline is therefore considered to be one of the oldest fishing methods, and handlining (jigging and trolling) is still widely used by both recreational and commercial fishermen.

Compared with hook and line or handline fishing, longlining is believed to be a more recent development and was not applied on a large scale until industrial production made hooks available at a reasonable cost. This did not happen until the last part of the nineteenth century, when the first 'automatic hook machines' were constructed. Prior to that it is reasonable to assume that fish hooks were regarded as fairly costly items so that the relatively large number of hooks that are needed to make a longline were not easily available or affordable. In the period from the Bronze Age to industrialisation, fish hooks were made by blacksmiths, and in areas where the demand for fish hooks was high there were specialised hooksmiths. In Bergen (Norway) for instance, the hooksmiths were organised into a 'hook-makers guild'. The hooks and rope needed for making longlines must have been rather expensive, as eighteenth century records of longline fishing in Norway give descriptions of this new and effective gear being used not by local fishermen but by people from the city of Bergen participating in the Møre and Lofoten cod fisheries. The longline was accused of being too effective, and longline fishing was also criticised by the church because its use led to swearing, quarrels, fights and abuse of alcohol. As a consequence, after its first introduction the use of longline gear was forbidden during several periods, and there are numerous records of fishermen being prosecuted and arrested for illegal use of the gear.

Fig. 1.5 Mediterranean (Catalonian) longline gear (demersal and semipelagic), eighteenth century. (After Lleonart & Camarasa, 1987.)

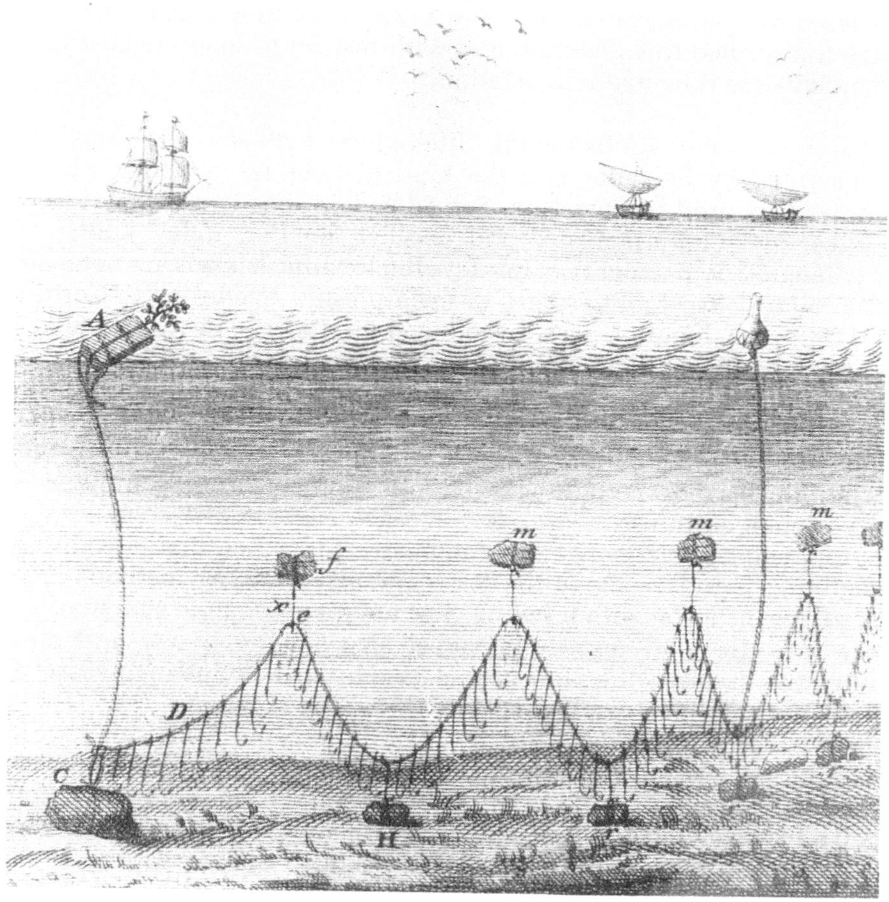

Fig. 1.5 *continued.*

This early resistance to the introduction of longlines may, in retrospect, be seen as a conflict between different social groups of fishermen rather than as resistance to the new gear based on rational arguments. This has also been the case when other new gears such as trawls or purse seines have been introduced in competition with the traditional fishing methods.

The use of longline gear in Norwegian waters can be dated back to the early 1700s. However, at that time Scandinavia was far more remote than it is today, so there are strong reasons for believing that this fishing technique was introduced from more central cultural areas. It may have originated in the Mediterranean region. A dictionary of Spanish fishing gear from the eighteenth century (by Sanez Reguart, 1791–95) includes descriptions and drawings of longline gear, both pelagic and bottom set (see Fig. 1.5). The original source of these drawings was information gathered by Joan Salvador i Riera in 1722. Lleonart & Camarasa (1987) have transcribed this material, and with respect to longline gear their comments are (Lleonart's translation):

> There are not fundamental differences between the longlines described by Salvador and the kind of longline 'a penjar' that is nowadays used in Catalonia. Salvador describes two types of longlines, one of normal size and another smaller, today known as 'palangret' or 'palangro'. Nowadays the longline baskets are not made by palm [leaves], but esparto or even plastic. According to Cornide (1788), the longline was introduced in Galicia by the Catalans, probably during the 18th century. The longline is cited by Despuig during the 16th century for the Tortose region.

Thus the use of longline gear dates back to at least the 1500s, but the technique has most certainly been applied long before this in the Latin and oriental cultures.

From being a relatively capital-intensive gear when first introduced, longlining is today among the less capitalised fishing methods compared, for instance, with trawling. The nature of longline gear makes it very versatile with respect to vessel size, engine power and crew number as basically the same gear is used by a small one-man vessel as by a large autoline vessel with 15 or more crewmembers.

Historically there have also been developments from small-scale, shore-based, coastal longlining to offshore large-scale and distant-water longline fishing. Early examples of this are the Greenland and Grand Banks fisheries for cod by Portuguese, Spanish and North American fishermen, with large sailing vessels as mother ships carrying 10–20 smaller dories that fished longlines with one or two crewmembers each. In the North Sea there was also a high longlining activity for cod conducted by sailing vessels. In the nineteenth century there was a large fleet of so-called cod smacks, sailing vessels operating from Lowestoft and other fishing ports of the English east and south coasts. Figure 1.6 shows a cod smack shooting the longline gear, perhaps at the Dogger Bank. This drawing is adopted from March (1953), which described the gear and fishing of the cod smacks:

> In the 1860's a complete set of longlines was made up of about 15 dozen, or 180 lines, each 40 fathoms in length, with 26 snoods 1.5 fathoms apart, with baited hooks attached. A 'string', made up of 180 lines, each fastened together, was about 7,200 fathoms long, or over 7 nautical miles, say 8 statute ones, with 4,680 hooks, each baited with a whelk. Soon after dawn, at about half-tide, the lines were shot across the tide to allow the snoods to drift clear of the main line, the smack being under easy sail, with the wind as free as possible so as to make a straight course while paying out.

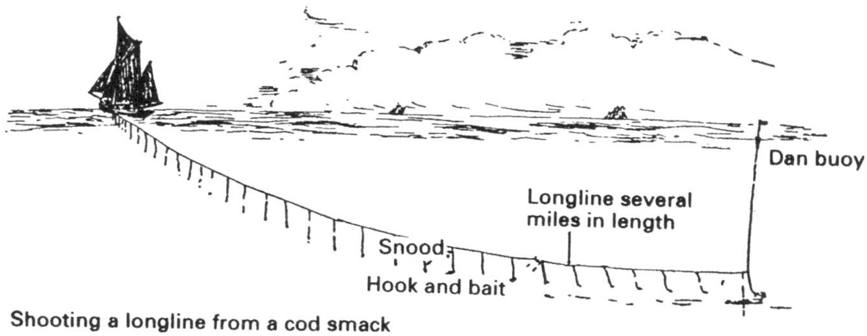

Fig. 1.6 Sailing smack shooting the longline gear in the North Sea. (Redrawn after March, 1953.)

However, further development of these offshore longline fisheries was hampered by the introduction of steam trawlers that totally outcompeted most other fishing methods and thereby changed the fishing pattern of offshore fishing. It is interesting to note that a similar shift from passive to active fishing gears did not happen in Norway, presumably partly because of lack of capital for investment in the necessary vessels and equipment but mainly as a result of strong resistance to trawling, both among fishermen in general and by the authorities. Although not so clearly outspoken, these attitudes have remained until recent times and are the main explanation for the varied pattern of different fishing gears being used in the Norwegian fisheries as compared with other North Sea countries where trawling is the dominant fishing method today.

Chapter 2
Principles of the Longline Catching Process

2.1 *Passive and active fishing gears*

Fishing gears are often classified into two main categories: passive (stationary or fixed) and active (mobile) gears. Trawls, seine nets and purse seines are the most important gears in the latter category, while gillnets, trammel nets, fyke nets, traps, pots, trolling, jigging, handlines and longlines are classified as passive gears. With active fishing gears the fish are chased by the moving gear; a groundfish trawl, for example, sweeps along the bottom and catches fish in its path. In contrast, capture of fish with passive gears is based on active movements by the animals towards the gear. With baited gears such as pots and longlines, this directed movement is induced and reinforced by smell stimuli from the bait that attract fish and cause them to congregate at the gear.

2.2 *Feeding behaviour and chemical sensing: the basis for fishing with baited gears*

Fishing with baited gears is based on the target species' demand for food. The bait is regarded by the fish as a food item, and fish are caught on longlines because the bait releases odours that are dispersed by the water current and trigger the fish to search for and ingest the baited hook. The behaviour of a fish responding to baited hooks is therefore similar to the behaviour of a foraging fish, and much has been and can still be learned about the longline catching process from studying the feeding behaviour of fish.

The visibility in water is limited, and food items can only be seen from short distances (a few metres). The chemical senses therefore play an important role in food finding and feeding in aquatic animals. Fish have well developed senses of smell (the olfactory system) and taste (the gustatory system). The olfactory and gustatory systems are both stimulated by various amino acids which comprise the most important group of compounds identified as feeding attractants and stimulants. The chemical senses in fish thus demonstrate some functional overlap, and are not as clearly distinguished as in terrestrial animals where the olfactory organ is stimulated by volatile airborne molecules and the gustatory system is stimulated through physical contact with the source of the stimulus.

However, smell and taste are clearly distinguished by the anatomy of their organs. The fish nose is a paired organ with an olfactory cavity or pit located on each side of the head. Each cavity has an anterior inlet and posterior outlet through which a flow of water brings the dissolved chemical compounds into contact with the olfactory receptors in the nasal cavity (Fig. 2.1). The olfactory organ enables the fish to smell feeding attractants released from food sources within a certain distance as these are dispersed by the water current.

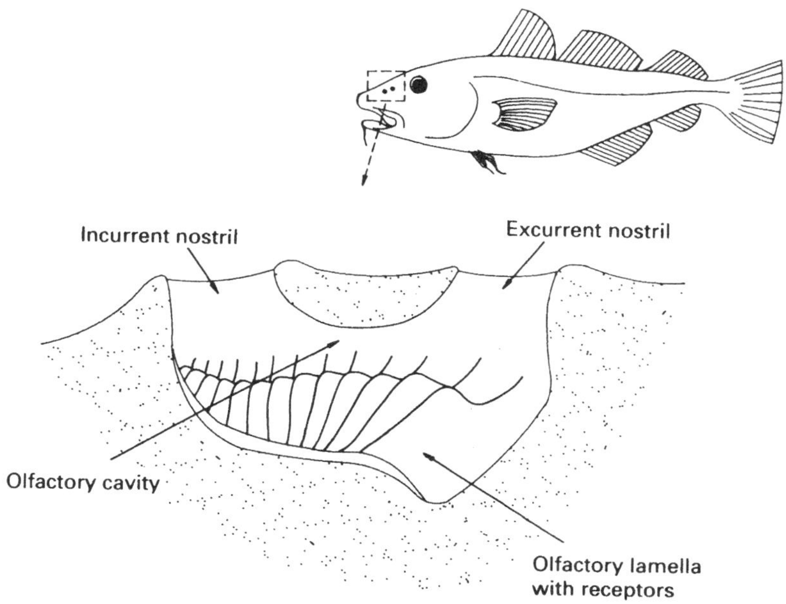

Fig. 2.1 Schematic drawing of the olfactory organ of a generalised fish. (Modified after Zeiske *et al.*, 1992.)

The gustatory receptor cells are gathered into structures called taste buds which constitute the basis for the gustatory organ. Taste buds are located in the mouth and gill cavities, and on appendages such as barbels and fins. Some species, such as the freshwater catfish, have taste buds over the entire body surface (Fig. 2.2). The catfish lives in waters of very low visibility. It has small eyes (poor vision) and therefore relies entirely on its well developed gustatory organ in food searching. Fish that have taste buds on barbels and fins (e.g. catfish, cod and hake) typically search for food by trailing their taste appendages along the bottom (Fig. 2.3).

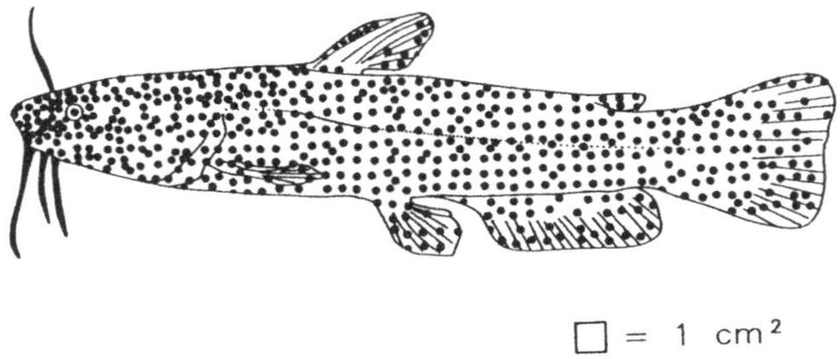

Fig. 2.2 Distribution of taste buds on the skin of an adult 27-cm catfish. Each dot represents 100 taste buds, i.e. number of dots per cm^2 = number of taste buds per mm^2. Solid black = more than 15 taste buds per mm^2. (Redrawn after Bardach & Villars, 1974.)

Fig. 2.3 Food searching behaviour of cod. When alerted by food odour, the fish is trailing the barbel and pelvic fins over the bottom to locate the odour source. (Drawing by Stein Mortensen.)

The chemical senses of fish are very sensitive and enable them to detect odour compounds at extremely low concentrations. Excretion products are continuously 'leaking' from aquatic animals, creating a 'smell trail' that can be utilised by a predator to locate its prey organisms. Different prey types release species-specific mixtures of chemical compounds that enable a fish to discriminate between different source identities: chemicals released from a shrimp differ from those released from a herring. The characteristics of a chemical signal received by the olfactory or gustatory organs are therefore important 'information for decision making', determining whether or not a fish will respond to the signal. As fishes are able to sense very low concentrations of chemicals, and also can distinguish between closely related chemical compounds, the chemical apparatus of fish is sensitive with respect to both quantity and quality of the chemical signal.

Chemical signals are not only used in food search. Prey organisms may use their chemical senses to recognise predators and escape. When returning to their native spawning sites for reproduction, salmon rely on the olfactory system to find and recognise their home rivers. Sex pheromones (specialised chemicals) released to the environment by females may elicit courtship behaviour in males, and parent fish of some species (such as cichlids) recognise the odour of their young.

2.3 *How fish locate baited longlines*

As visibility in water is limited, the chemical senses are most important when fish become alerted to baited hooks and are searching for their location. Hence, it is through stimulation of their chemical senses that fish typically become aware of the presence of baited longlines. Chemical stimuli, however, have no directional cue that can guide the fish to the bait. In this context they differ from visual and sound signals, which give information also about the location of the source of stimuli. This difference can be illustrated by the following example. If you are sitting in the kitchen when somebody is making French fries, you can see and hear where the cook is working. However, the smell from the cooking does not tell you anything about the cook's location in the room. Similarly, the fish must therefore use another cue when searching for an odour source.

This cue is the current direction. Attractants released from the baits are dispersed by the water current in the same way as the wind transports scents in the air (Fig. 2.4). A deer will become aware of a hunter who approaches the animal from the upwind direction. Similarly, it is only the fish that are located downstream from a baited longline that can sense the bait odour and locate the baited hooks by swimming upstream. Underwater video cameras have been used to study responses of fish to fishing gears, and for both baited hooks and pots it has been shown that most fish approach the gear by swimming upstream (a process termed *chemically stimulated rheotaxis*).

This way of approaching and locating baited gears may explain why the catch rates of such fishing methods are affected by the velocity of the current. As fish predominantly swim upstream towards a baited longline, it would be energetically advantageous to stay in shelter when the current is strong, and to be active during periods with moderate or low current velocities. Longline fishermen therefore often experience low catch rates during periods of strong currents, which may be explained by the lower swimming and feeding activity in these periods.

At a certain distance downstream from the longline, the concentration of attractants released by the baits matches the chemical sensing threshold of the fish, and beyond this distance the bait scent has no further effect. This distance over which fish would become alerted to and exhibit a searching response to baited longlines, i.e. the spatial extent of the area influenced by the gear (often called 'the active space'),

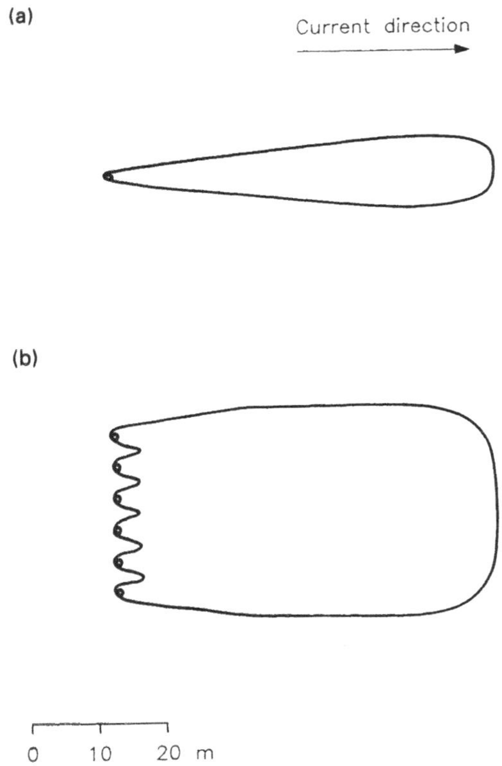

Fig. 2.4 The shape of an odour plume after a 2-h soak time at a current speed of 0.6 cm/s. (a) The odour plume of a single bait, and (b) the odour plume of six longline baits. (Redrawn after Olsen & Laevastu, 1983a.)

is of interest to the fisherman. When he sets his longlines in parallel fleets, he wants to spread the fleets so that the spacing is larger than the area influenced by the bait odour. With shorter spacing between the longline fleets, they will attract fish from overlapping areas so that the total extent of the fished area will be smaller, giving suboptimal catch rates.

The length of the odour plume within which fish would exhibit food searching behaviour is determined by the extent of the area where feeding attractants released from the baits are present in quantities that are high enough for the fish to smell them. The distribution of this area is affected by the following factors:

(1) The concentration of attractants released from the bait, i.e. the rate of chemical release. Higher release rates give rise to large odour plumes.
(2) The rate of dilution of the attractants as they are dispersed by the current. In turbulent environments, the attractants are more dis-

persed horizontally and vertically, causing the odour plume to be shorter.

(3) The capability of the fish to smell the attractants, i.e. its sensitivity to bait odour. Fish that have a high sensitivity to smell stimuli are able to detect the attractants at lower concentrations, and the odour plume will therefore be larger for more smell-sensitive species.

The attractant release rate from a bait is initially high when the bait is submerged in seawater, but decreases rapidly (Fig. 2.5). Different scientific methods have been applied to determine the sensitivity of fish to bait odour, and one way of doing this is illustrated in Fig. 2.6. Calculations of

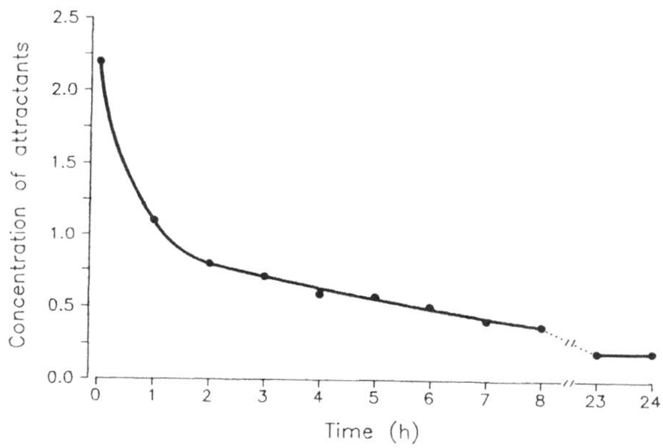

Fig. 2.5 Rate of release of attractants (amino acids) from mackerel bait. The concentration of attractants is given as μmole alanine-units per minute. (Redrawn after Løkkeborg, 1990a.)

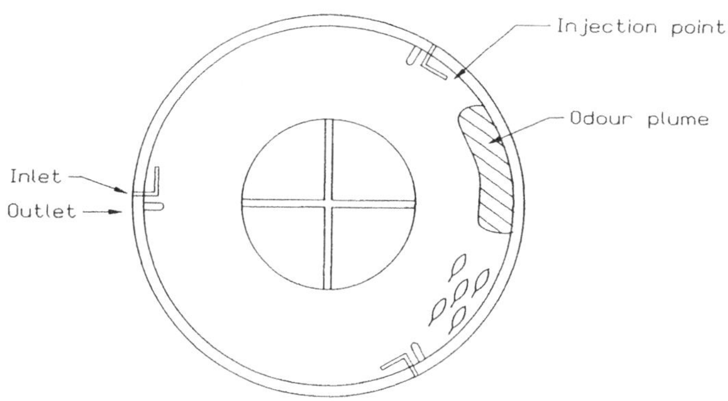

Fig. 2.6 Experimental pool used to determine the sensitivity of sablefish to bait odour. Sablefish responded to bait odour injected into the pool by turning repeatedly within the odour plume. Their response threshold could be determined by successively injecting bait odour of decreasing concentration until no response was observed. (Redrawn after Løkkeborg *et al.*, 1995.)

the lengths of the active space can therefore be made by using these values of release rate and sensitivity in models that simulate the dispersion of odour plumes. Such estimates indicate that the active space for sablefish, which is caught by longlines and traps in the northern Pacific, may range up to several kilometres for food-deprived fish and with relatively high current velocities. With increased feeding motivation due to food deprivation, fish would respond to more diluted bait odours, and the dilution of the odour plume due to turbulence is lower in stronger current. These estimates of the length of the active space are based on theoretical models of odour dispersion that have several assumptions, and they have not yet been verified by field observations.

However, tracking studies have confirmed that Atlantic cod responds to baited longlines from a distance of at least several hundred metres. By tagging the fish with ultrasonic transmitters their movements can be tracked. This technique has been used to study the responses of cod to baited longlines. When a baited longline was set upstream of the tagged fish, they responded by swimming towards the line and were caught (Fig. 2.7). Such upstream (rheotactic) orientation to baited hooks was

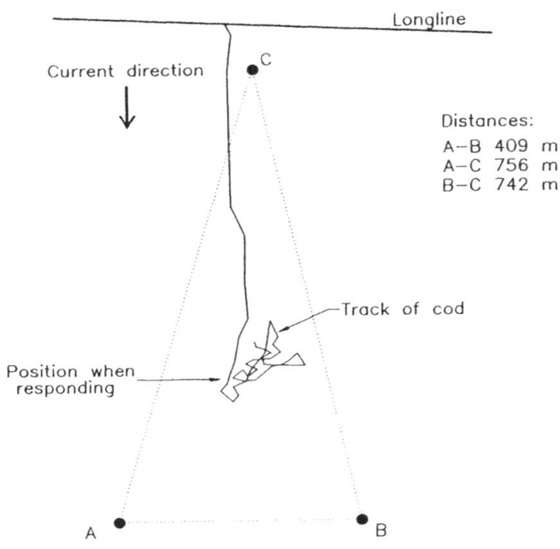

Fig. 2.7 Track of a 51.5-cm cod responding to and locating a baited longline. A, B and C show the positions of the hydrophone buoys of the positioning system used to track the cod tagged with an ultrasonic transmitter. The track shows that the fish was milling around and often changed swimming direction for a period of 4 h 5 min after the longline was set. Then, from a distance of 575 m, the fish changed swimming pattern and swam straight towards the longline and was hooked. The current speed in the area was measured at about 4 cm/s, which indicated that the fish responded at the time the odour plume reached its position.

demonstrated for fish as far as 700 m from the longline. With weak near-bottom current of only a few centimetres per second, the odour plume disperses very slowly and does not reach fish at this distance for several hours. Thus, the longline must be soaked for a relatively long period in order to attract and catch fish from the far field.

As current speed is significantly reduced close to the bottom, the spread of bait odour from bottom set longlines is thought to be relatively slow compared with longlines set higher in the water column. Although the effect of increasing current speed with increasing height over the bottom has not been investigated, a possible larger active smell space could be an additional reason why semipelagic longlines are often found to be more effective than bottom set gear.

Although visibility in water is limited, vision may play an important role in food location when light intensities are sufficient. In fishing with pelagic longlines, vision may be of importance in locating the baited hooks because baits in midwater are more readily seen than baits on the bottom. Tunas constantly cruise in the upper water layers in their search for prey, and use vision when attacking the prey. The cruising behaviour of foraging tuna may bring fish into contact with the baited gear even if the fish are not triggered by the bait odour, and the baited hooks may therefore be located by means of visual cues. Similarly, cod are caught by semipelagic longlines when they are migrating to the spawning grounds, and in this fishery vision may be important in bait location as the migration brings fish into contact with the baited gear.

2.4 The hooking process

Once the longline has been located, the fish decides whether to accept or reject the baited hook as a food item. In deciding whether to attack and ingest the baited hook, the fish judges visual, chemical and mechanical aspects of the baited gear. The visibility of the gear (snood and mainline) and the size and shape of the bait are factors of importance in this phase of the hooking process. Monofilament lines have proven to be far more effective than multifilaments lines, particularly under good light conditions where they have been observed to catch as much as three times more cod than multifilament lines. This is probably due to the higher visibility of multifilament lines, which may cause a restrained response towards attacking the baited hooks. Also gillnets made from different types of material have different catching efficiencies. Monofilament gillnets are more efficient than the multifilament type which the fish can see more easily and therefore react and turn away before coming into contact with the gillnet. The possible positive effect of the lower visibility of monofilament lines on catch rates has not, however, been proven. A series of underwater observations in the field on the contrary indicate that longline target species (which, with very few exceptions, are predatory species) show little or no fright reactions when approaching the gear, be it multifilament or monofilament. Another explanation for the difference in catching power between multi- and monofilament gear could be the higher affinity of multifilament gear to absorb smell stimuli from the bait, so that after some time of use the whole gear (mainline and snoods) is 'contaminated' with smell stimuli, in contrast to monofilament, which has a relatively small and smooth (non-absorbing) surface area. Fish approaching a multifilament line might therefore need a longer time to locate the bait, as smell stimuli might be distributed along the complete gear. With monofilament lines the baited hooks might appear as more distinct objects of smell and therefore more easily distinguishable for the approaching fish. Alternatively, the smell stimuli absorbed by multifilament lines may after some time create a rotting scent, repelling the fish.

The visibility of the bait may be important in attracting fish and trigger them to attack the bait. Both coating natural bait with luminous bacteria and using fluorescent artificial bait in addition to natural bait have given rise to a pronounced increase in catch rates in Japanese experiments with bottom longlines. The catching efficiency of baits such as mackerel, herring and sardine may partly be due to their light-reflecting skin, which makes these baits more visible.

The size and shape of the bait also affect whether the bait is attacked or rejected. Fish show preferences for prey of certain sizes; larger fish prefer larger prey, and the prey size of small fish is limited by their mouth size and their capability to catch and handle the prey. Therefore, fewer smaller fish respond to large than to small baits, and this behaviour difference has important implications for the size selectivity of longlines (see Section 6.3). Only a small proportion of the fish that are attracted to a longline do actually attack the baits (see below). A bait such as a piece of mackerel or squid represents a novel food item that the fish most probably has never experienced before. This may cause a somewhat restrained response towards the bait and explain why there is a fairly low proportion of attracted fish that apparently attack the baited hooks. This proportion is probably lower for smaller fish, because they have a narrower diet breadth and are subject to a higher predation risk than larger individuals (i.e. they have more enemies) and therefore show a more restrained response towards novel prey items.

Before ingesting the bait, the fish uses its taste buds on the barbels and lips and in the mouth cavity to judge the taste of the bait. In this phase of the hooking process, the fish also tests the bait texture using its mechanoreceptors that lie close to the taste buds. Before ingesting food some fishes use both taste and texture to sort food items from less palatable or inedible material. Taste is probably the most important factor for bait acceptance. The fisherman is aware of the importance of bait type and bait quality when choosing bait. The choice of bait is determined according to the target species because different species are caught most efficiently by different bait types. Thus the species-specific

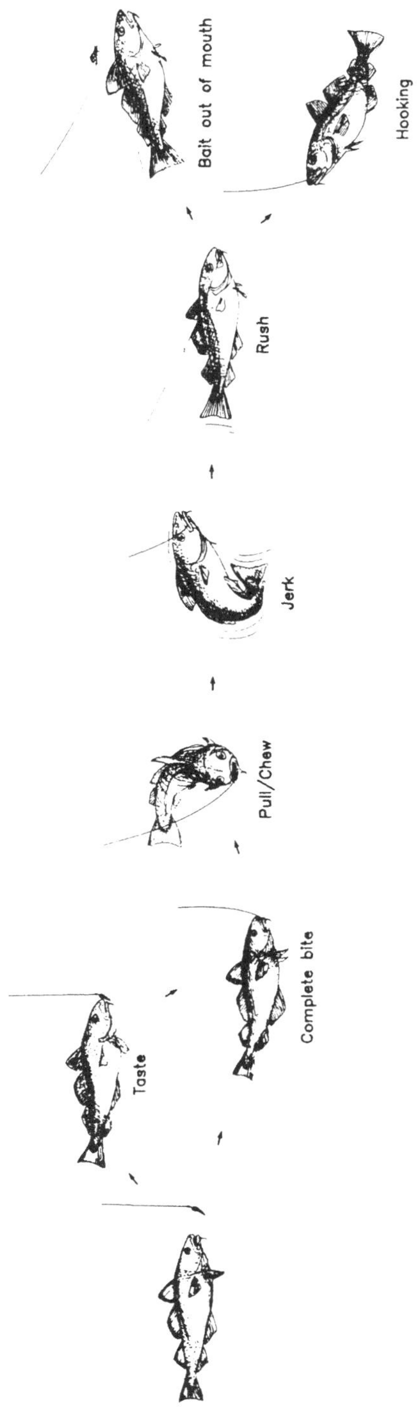

Fig. 2.8 The hooking behaviour of cod. (Redrawn after Løkkeborg *et al.*, 1989.)

preference for different bait types affects the species selectivity of longlining (see Section 6.2).

Even though the hooking behaviour differs between species, we can use the typical behaviour of cod to illustrate the hooking process in longlining (Fig. 2.8). An attack by cod towards the baited hook is usually initiated by biting the bait. Sometimes a brief taste by touching the bait with the lips or the barbel is made before the fish decides whether to ingest the bait or not. After biting the bait, the fish makes a rush (swimming rapidly forward) or a few jerks (rapid sideways movements of the head). The attacking sequence is terminated with hooking or the baited hook being pulled out of the mouth. Hooking is thus a consequence of the fish making a rush or jerk that stretches the snood and causes the hook point to penetrate the tissue of the mouth cavity or the alimentary tract if the baited hook has been swallowed.

When a fish is hooked, it normally fights vigorously for a short period (about a minute), trying to escape, and then comes to a rest. However, during the remaining soak time of the gear, the hooked fish will occasionally repeat the escape behaviour, spending most of the time resting on the bottom (from observations mainly done with demersal longlines). On retrieval of the gear the fish will initially perform escape movements as it is being lifted by the line. Later during the hauling of the line the fish is struggling less and is mostly being passively hauled through the water. Some species, for instance the tusk, tend to curl the tail and body, which makes the fish rotate continuously as it is pulled by the line. This often causes the snood to be severely twisted and shortened and/or tangled around the mainline, which increases the risk of fish loss during hauling (see Section 5.5).

Underwater observations have shown that only a small proportion of the fish that are triggered by the bait odour and locate the gear do actually attack the baited hooks (Fig. 2.9). Behaviour observations of haddock responding to baited hooks have shown that 36% of the total number of fish observed in the vicinity of the line made an attack towards the bait (approached, touched or bit the bait), and of these only

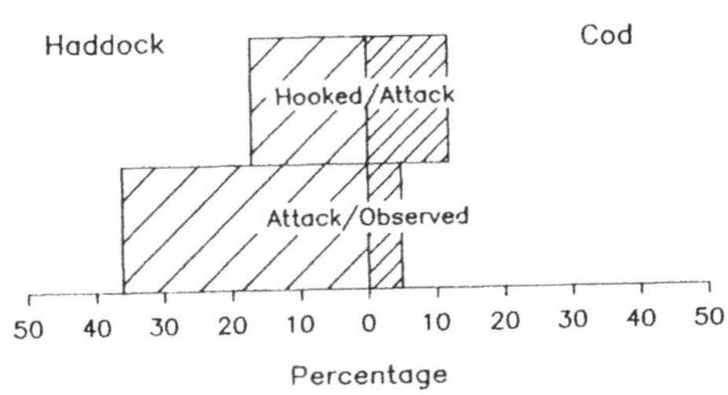

Fig. 2.9 Reactions of haddock and cod to mackerel-baited hooks. The proportions of observed fish attacking the bait, and the proportions of attacking fish being hooked. (Modified after Løkkeborg *et al.*, 1989.)

17% were hooked, i.e. only 6% of the haddock observed in the vicinity of the baited line were eventually hooked. The proportion for cod was even lower with less than 1% of the fish being hooked (5% attacked the bait and of these 12% were hooked). These observations indicate that a fairly low fraction of the fish available to longline gear is caught, and further that longlines have a higher catching power for haddock than for cod. Thus when fishing in an area where haddock and cod are equally abundant, longlines will catch much more haddock than cod.

Trawl is regarded to be about equally effective for these two species. When fishing with trawl and longline in the same area, the proportion of haddock has been observed to be about six times higher in longline catches than in trawl catches (Fig. 2.10).

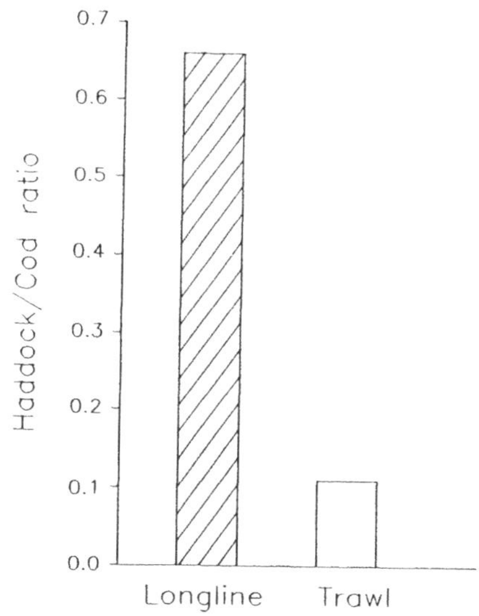

Fig. 2.10 Comparison of the ratios of haddock and cod caught by longline and trawl, based on catch data from a longliner and trawler fishing simultaneously at the same fishing ground in the Barents Sea. (Redrawn after Engås *et al.*, 1996.)

Chapter 3
Gear Construction, Operation and Vessels

3.1 *The gear: mainline, snoods, materials, hooks, hook spacing and bait*

There are numerous ways of rigging longlines according to target species and fishing conditions. Some typical longline riggings are shown in Fig. 3.1. The traditional bottom longline for species such as cod, haddock, ling and tusk has a multifilament mainline and multifilament snoods, and in recent years it has become common for it to be rigged with swivels and EZ-baiter hooks, with a hook spacing of 1.2–1.8 m. Semipelagic longlines for cod and haddock have monofilament mainlines and snoods with swivels and J- or Wide Gap hooks with 2–3 m hook spacing. Tuna longlines typically have a multifilament mainline to which the snoods are attached with a metal snap at wide hook spacings (about 50 m). The relatively long snood often has a multifilament part closest to the mainline and a monofilament part closest to the hook – usually with a swivel connection between the two parts. The diagram in Fig. 3.1(c) shows one simple monofilament tuna longline snood (right) and a more sophisticated snood configuration (left). Closest to the snap the latter snood has a 12-m-long, 5-mm three-strand multifilament line, and then, in sequence, a monofilament line (3 mm, 13 m), a swivel, a coated steel wire (3 mm, 2 m) and then another monofilament line (2.5 mm, 13 m) closest to the hook.

The mainline

The mainline is characterised by the material, the material construction and the dimension or thickness, normally given as the diameter (mm). As for most other fishing gears, synthetic fibres have replaced natural

---

Fig. 3.1 (Opposite and over page) Different longline riggings. (a) Typical traditional and modern bottom set longline with swivel, e.g. for cod fishing; (b) semipelagic monofilament longline for cod fishing; (c) pelagic longline for tuna. (Diagram from Fishing Tools Catalogue, Yoshimura Co. Ltd, Japan.)

(a)

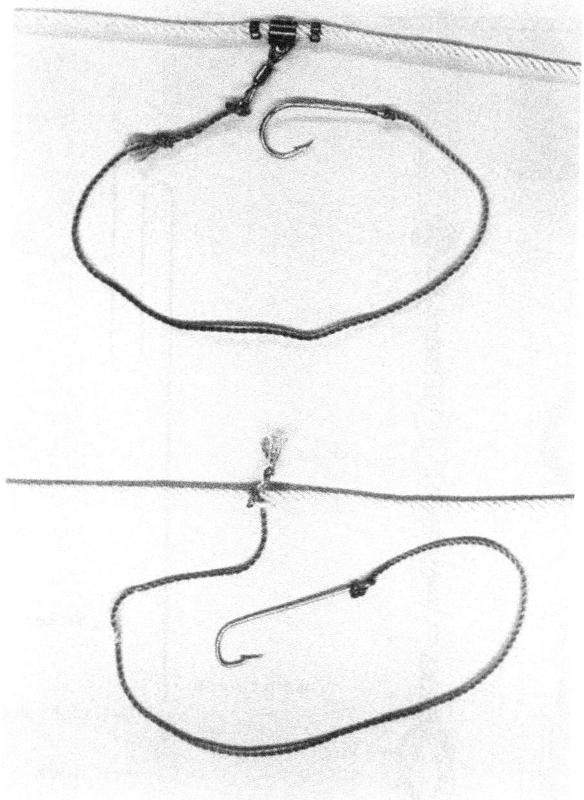

(b)

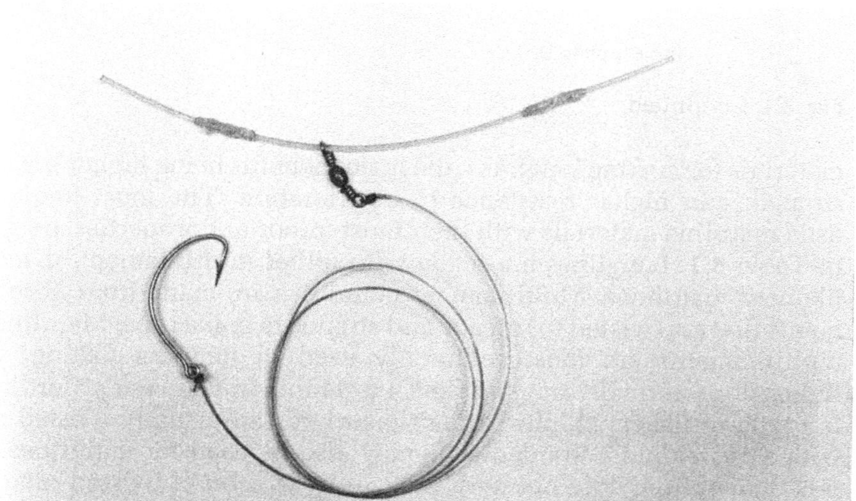

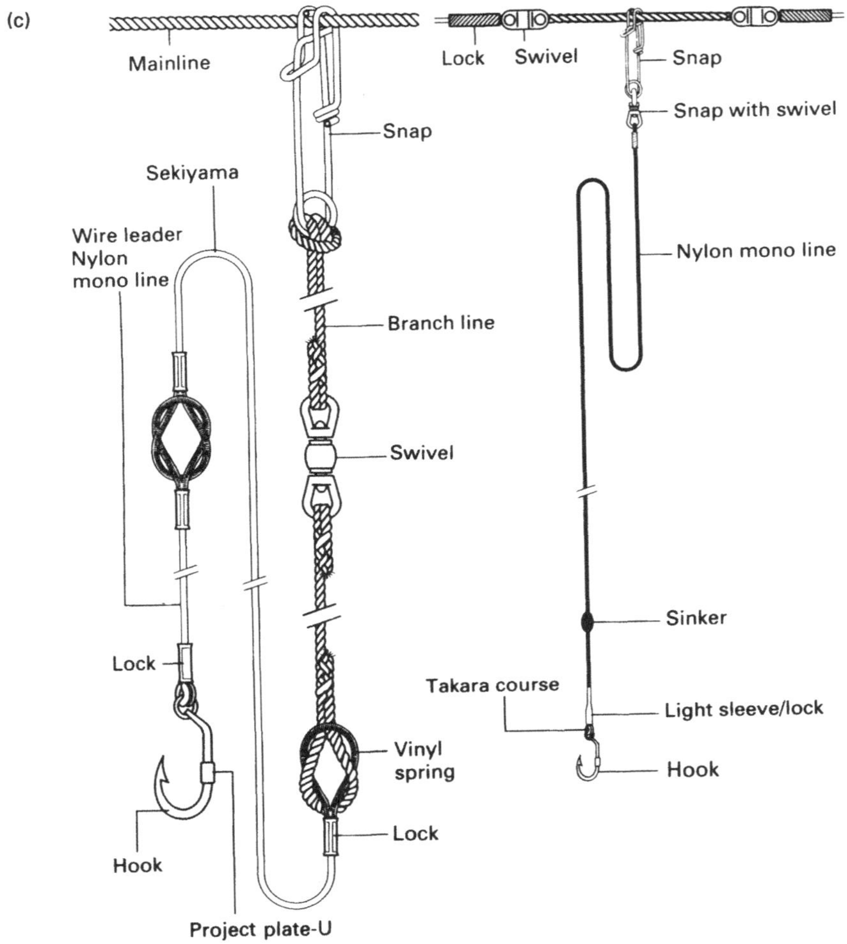

Fig. 3.1 Continued.

materials for making longlines, the major benefits being higher breaking strength and higher resistance to deterioration. The most frequently used mainline materials with their most important properties are given in Table 3.1. Longlines are made with either multifilament or monofilament mainlines. Multifilament mainlines are made from fibre filaments that are twisted to threads and strands to make rope. Mainlines of multifilaments are most commonly used in longline fishing, with dimensions normally ranging from 4 to 11 mm in diameter according to the type of fishery. Multifilament mainlines are usually twisted rope with three strands. Braided rope may also be used for mainlines, but this type of rope does not have the coiling tension of twisted rope and the latter type is therefore preferred because of its better coiling properties, an important feature of mainlines. Loose lines that do not coil

Table 3.1 Mainline materials and their properties

	Polyester	Polyamide (nylon)	Polypropylene
Breaking strength (g/denier)	6–7	7–9	8–8.5
Breaking elongation (%)	6–14	12–18	18–22
Specific gravity	1.38	1.14	0.91

properly are easily tangled, which complicates the gear handling both with tub-based and mechanised longlining.

During the last 30 years or so, monofilament mainlines have, however, gained increased popularity in longlining, mainly because the catching performance of monofilament lines has been shown to be superior to that of multifilament lines. In contrast to the multifilament line that is composed of a large number of thin fibres, the monofilament line, as the name suggests, has one filament only, commonly made from polyamide (nylon). However, their lower breaking strength and poor resistance to chafing mean that monofilament longlines can seldom be laid directly on the bottom except in areas with a very smooth sea bed (sand or clay). Monofilament mainlines are therefore mainly used in pelagic or semipelagic longlining.

High breaking strength and good resistance to chafing are mentioned as important properties of mainlines as is the stiffness of the rope. The last is related to the so-called coiling properties of the line because a prerequisite for effective handling of longline gear is that the mainline coils nicely with a low risk of tangling. For multifilament longlines the rope should therefore be made with a medium to hard twist to obtain the proper stiffness. Multifilament longlines are normally treated with coaltar or some other impregnating material, to improve both the handling properties and the lifetime of the line. To obtain good coiling and handling properties of monofilament mainlines, the monofilament made for longline purposes is given a certain heat treatment. This affects the molecular structure of the material, so that the line keeps its coiling properties even after being exposed to strong stretching during hauling of the line.

Snoods

As in the case of the mainline, we find longline snoods made from either multifilament or monofilament. However, in some longline fisheries, particularly for different sharks, snoods of steel wire or chain are used (Fig. 3.2) since fibre snoods are easily cut by the sharp teeth of these fishes. Monofilament snoods range from 0.3 to 1 mm (dia.) in thickness and from 0.5 to several metres in length according to the type

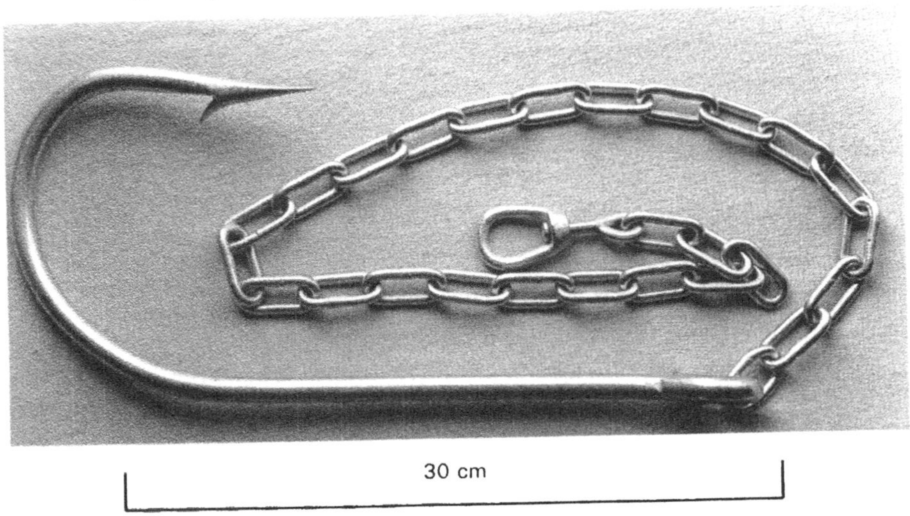

Fig. 3.2 Shark hook with snood of steel chain.

of fishery. Multifilament snoods may be either twisted (normally two-strand rope) or braided with thicknesses from 2 to 4 mm (dia.) and with lengths from 0.3 to several metres.

Hooks

Although attempts have been made to develop fish hooks from plastic materials, today they are made exclusively from metal (steel). The manufacture of a hook starts with a relatively thick steel thread, which is stretched to obtain the desired thickness for production of a specific hook size. Then the steel thread is cut into appropriate lengths, and through several steps it is made into the final product: a functional fish hook (Fig. 3.3(a)). A standard fish hook is an apparently simple item, with a shank, bend, point, barb and an eye or plate for attaching the snood (Fig. 3.3(b)). There are, however, an immense number of fish-hook varieties available from different hook manufacturers. A qualified guess would be about 50 000 different hook designs, but this number may be significantly larger so there are far more hook varieties than fish species in the world. The largest variety, however, is found in hooks for sport fishing. So-called commercial hooks (hooks used for commercial fishing) have a more restricted diversity, but each variety is normally made in large quantities. The great diversity in hook types, which is mainly a result of the demands of different customers, is made possible by the properties of the material (steel) and the techniques of manufacture. The main variables of a fish hook are size, shape and coating,

(a)

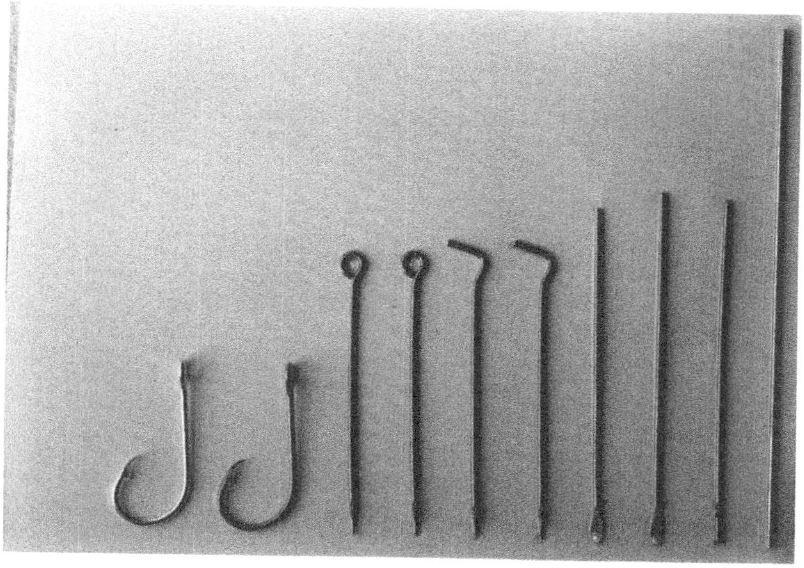

(b)

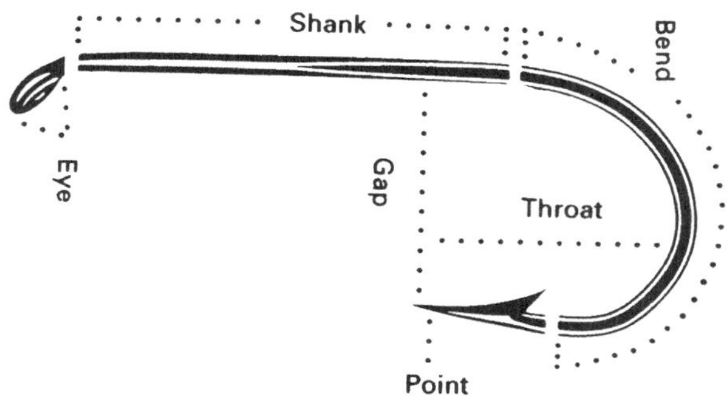

Fig. 3.3 (a) Fish hook manufacture: the steps of automated production from steel thread to a functional fish hook. (b) Fish hook: basic parts. (Redrawn after hook catalogue of O. Mustad & Son Ltd.)

which give an indefinite number of possible combinations. Fish-hook terminology is a somewhat confusing subject, since there is no strict international standardisation of definitions. Here the terminology of the hook manufacturer O. Mustad & Son Ltd is used to illustrate the main terminology, and to a large degree this is also used by other hook manufacturers. Figure 3.4 shows a typical label from a box of longline hooks, with a description of the main characteristics of the hook: hook name, quality number, special features (ringed, point curved in), coating (special Duratin) and size (13/0). These are the main characteristics of the hook sold under that specific label, and should always be quoted when specifying the size and shape of that particular fish hook.

However, the total diversity of hook parameters is much greater than

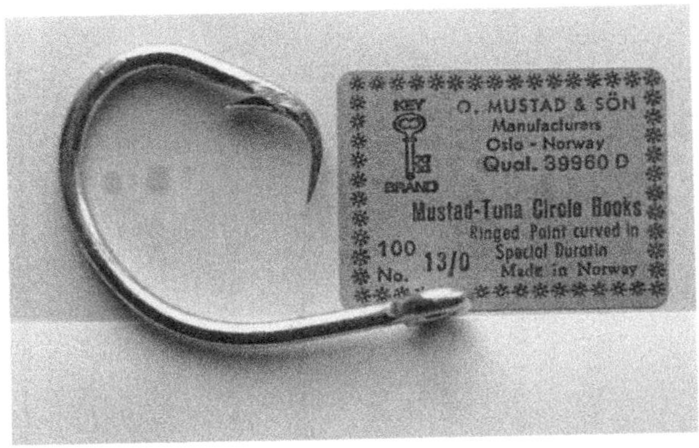

Fig. 3.4 Typical label from a box of longline hooks.

those given for one specific hook type. To give a comprehensive explanation of the classification of fish hooks is beyond the scope of this book, but some further explanation of different hook parameters is needed.

Hook size

The size of a fish hook may vary considerably. However, most longline hooks are of medium to large size. The hook size is indicated by a number. Hook size numbering might seem confusing, but as a rule hook size decreases with increasing ordinary numbers (1, 2, 3...) and increases with increasing '/0' numbers (1/0, 2/0, 3/0...), as shown in Fig. 3.5, where this is illustrated by the size range of the Mustad–Viking hook, Quality No. 79 568.

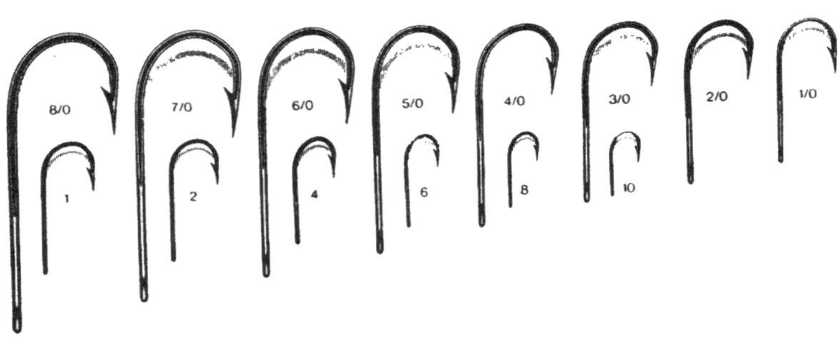

Fig. 3.5 Hook size numbering (not to scale). (Redrawn after hook catalogue of O. Mustad & Son Ltd.)

Hook shape

The name of the hook (Circle, E-Z Baiter, Kirby, O'Shaughnessy, Wide Gap) indicates its basic shape and the quality number specifies varieties within this hook 'family'.

Major variations within the basic shape of a hook and other special features are shown in Fig. 3.6. A hook can be straight or with the point turned left (kirbed) or right (reversed), features that triple the number of varieties within a certain hook type. The shank may be 'forged' which makes the hook stronger and more resistant to forces acting in the hook plane. In addition there is a series of different special hook features related to the configurations of the point and shank termination, which add considerably to the number of possible hook varieties.

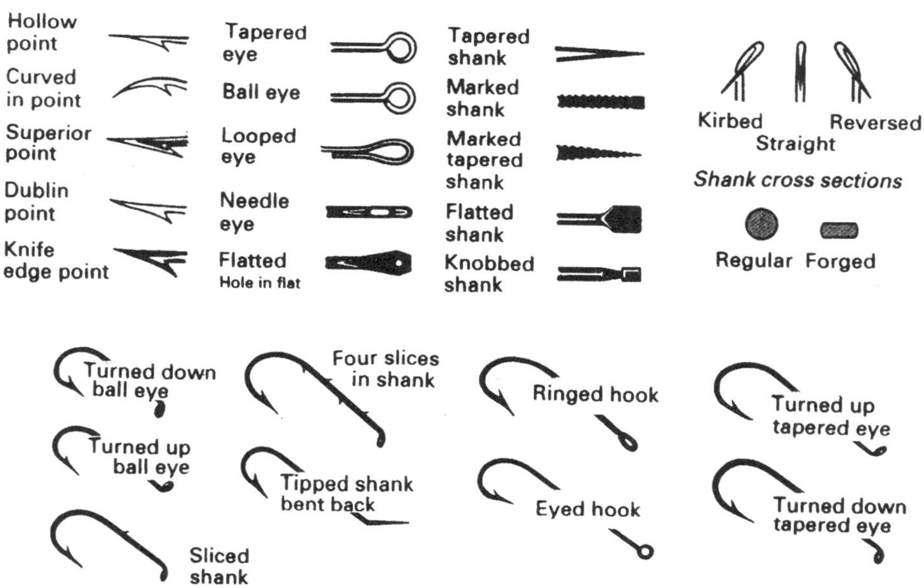

Fig. 3.6 Terminology of special fish hook features. (Redrawn after hook catalogue of O. Mustad & Son Ltd.)

Coating

To make the hook corrosion resistant, it is coated using electrolysis with different metals such as tin, nickel, cadmium or combinations of these or other anti-corrosives.

Considering the possible combinations of size, shape and coating, it is easy to imagine the large variety of hook types available. However, as mentioned above, the number of different hook types used in commercial longlining is fairly restricted. In summary the hook should have maximum catching efficiency, it should be easy to work with in handling and baiting, and it must fulfil the requirements for strength and elasticity. This is obtained in the hardening process, which is critical to make the hooks neither too soft nor too brittle.

Fish hooks for commercial longlining are supplied by a number of manufacturers, large and small. In the Asian market (e.g. Japan), hooks for small-scale longlining may still be supplied by various local hookmakers whereas the large scale and western longline fisheries are mostly supplied with hooks from a few, larger, industrialised hook manufacturing companies. Of these, O. Mustad & Son Ltd of Norway is by far the largest. Other main suppliers of longline hooks are Wright and McGill (Eagle Claw) of the USA, Milward's Fish Hooks of England, and VMC Pêche of France.

Hook spacing

The hook spacing also varies between different longline fisheries. In cod fishing the common hook spacing is 1–2 m, while in tuna fishing it may be as much as 50 m.

Baits

A large variety of bait types are used in longline fishing, for example herring, mackerel, anchovy, sardine, sprat, pilchard, saury, saithe, capelin, shrimp, crab, squid, octopus, various mussels, whelks and lugworms. The bait organisms may be either used whole or cut into adequate pieces (Fig. 3.7). The combination of two types of bait on the

Fig. 3.7 Baited hook, with whole sardine and piece of mackerel. (Redrawn after Franco *et al.*, 1987.)

same hook, as shown in Fig. 3.7, has been tried in experimental longlining. However, such baiting arrangements are rarely found in commercial longlining. When combinations of different baits are used in regular longline fishing, the different bait types are applied alternately along the line, e.g. ten hooks with squid, then 20 hooks with mackerel and so on. The choice of bait in different longline fisheries depends on its catching efficiency for the target species, but also on the availability, cost and handling properties of the bait.

3.2 Methods of setting longlines

Longlines are set in different ways, but there are three main methods of setting:

- demersal
- semipelagic
- pelagic longlining

Demersal longlining

Demersal longlining (bottom set gear) is the traditional and still most common method of setting, and is used for demersal species such as cod, halibut, hake and ling. Fishing is most often carried out at depths from 100 to 800 m, but in some fisheries even down to 2500 m (as practised in the fishery for Patagonian toothfish off Argentina). The longline fleets are laid on the sea bed with an anchor, buoy line, buoy(s) and/or marker buoy at either end (Fig. 3.8). Intermediate buoys with buoy lines and anchors may also be used, for example, for every five

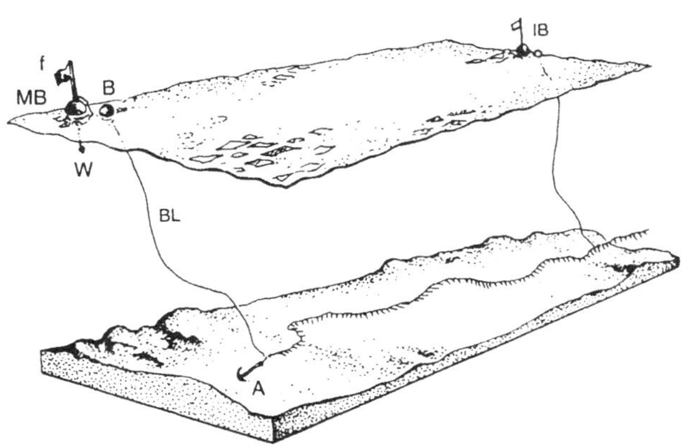

Fig. 3.8 Demersal (bottom set) longline with anchor (A), buoy line (BL), buoys (B), intermediate buoy (IB) and marker buoy (MB) with weight (W), pole and flag (f). (Drawing by Stein Mortensen.)

skates, to save time and to minimise the risk of losing gear when the mainline breaks occasionally during hauling. Instead of steaming to the end buoy, which might be several nautical miles away, the closest intermediate buoy is picked up and the hauling continues after a short break. As indicated by the name, the main function of the marker buoy is to mark the ends of the longline fleet so that it can be easily spotted from a distance. It consists of a 3–4 m high pole mounted on a float and with a weight at the bottom end to keep it floating in an upright position. To make it more visible it often carries one or two flags at the top end and normally also a battery-and-light package for easy identification in darkness. The marker buoy may also have radar and light reflectors for easy identification by searchlight and radar. In addition to the marker buoy there are normally one or more surface floats, the main function of which is to take the strain from the buoy line. In strong tides one or more of these floats may be pulled under the surface by the current force acting on the buoy line. Among longline fishermen the tidal flow is commonly stated as 'a quarter, one or one and a half float of current'. Marker buoy and floats also normally carry the identification number and name of the vessel. The buoy line is usually a rope somewhat stronger than the mainline, because the buoy line must take the high strain that is often needed to break loose the anchor.

The anchor (or grapnel), the function of which is to keep the longline fleet in position, is normally made of steel with three or four claws to get a good grip in the bottom substrate. The anchor may weigh from a few kilograms to about 80 kg, according to fishing depth and tidal conditions. A second function of the anchor is to make the gear sink faster to the bottom with increasing depth and tidal flow, so significant displacement of the line due to drift during sinking is avoided with the use of heavy anchors. For the same purpose, smaller weights like stones, steel rings or steel chains (1–5 kg) are attached to the mainline, often at the link between skates. In small-scale longlining, weights of stone or some other heavy material are still widely used as end weights instead of regular anchors.

How the total amount of gear is operated by a longline vessel depends on the bottom topography and the distribution and density of fish (Fig. 3.9). With high fish densities over a large area it is common practice to set all the gear (15 000–30 000 hooks) in one long fleet or 'long-setting', either in a fairly straight line or in a U-form, e.g. in the longlining for cod on the banks off northern Norway. When fishing the continental slopes for species such as ling and tusk, the gear is divided into smaller fleets of 1500 to 3000 hooks that are set parallel along the slope, with a shallow end (e.g. 200 m) and a deep end (e.g. 500–600 m) of the fleets. Where the fish density is low, the fleets are spread fairly wide apart, often around 800–1000 m. As the first fleets are hauled they are moved and set at the 'front' of the fleet system. Consequently, the longliner will fish the slope in one direction, rather similar to the ploughing of a field.

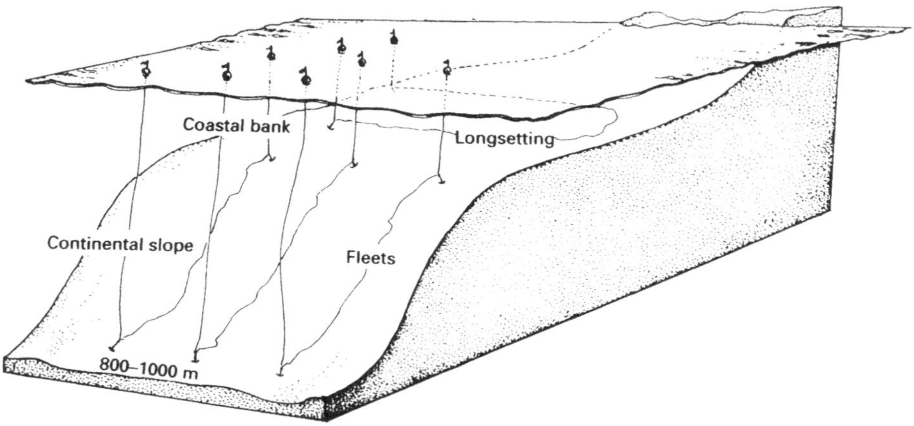

Fig. 3.9 Modes of fishing bottom set longlines. (Drawing by Stein Mortensen.)

However, if very high catch rates are obtained, the fleets are reset in the same position and a vessel may fish in a fairly small area for days or even weeks as long as the catch rates remain good.

These are the main modes of operating bottom set longlines, but there are of course a large variety of adaptations depending on local conditions. One of these is fishing on very concentrated fish aggregations which often occur on undersea mountain peaks, where the success of fishing with longlines depends on techniques for deploying a large number of hooks over a small area. For instance, off the Portuguese coast this has led to the development of a method where the longlines are set in an overlapping criss-cross pattern on narrow sea mounts. This leads to a lot of line tangling, so that the lines often have to be totally re-rigged before the next setting.

Semipelagic longlining

Semipelagic longlining includes elements of bottom set longlining and (true) pelagic longlining. As for bottom set longlines, the fleets of gear are kept in a fixed fishing position by anchors while the mainline is set pelagically at variable fishing depth as for pelagic or drift longlining. Two ways of semipelagic longlining are illustrated in Fig. 3.10.

Semipelagic setting of longlines is aimed at fishing effectively with longlines when the fish are distributed at various depths in the water column relative to the bottom topography. As the depths of fish concentrations are registered by echo sounder, the mainline is equipped with floats and sinkers at intervals in order to match fish depth and gear depth as closely as possible. One way to obtain this is to attach floats and a float/sinker/sinkerline arrangement alternately along the main-

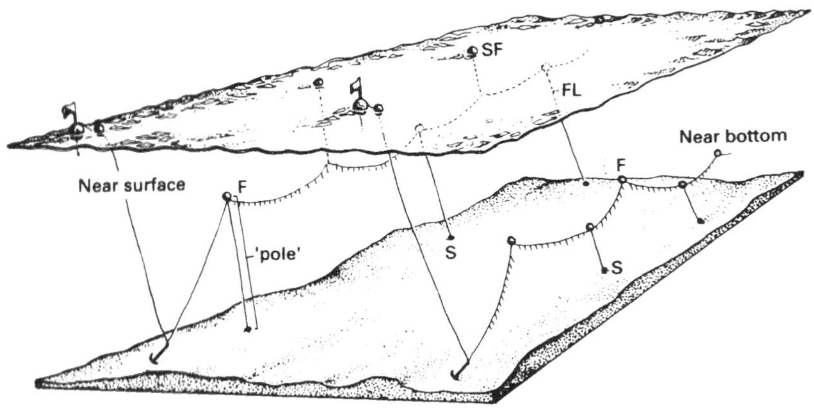

Fig. 3.10 Semipelagic longlining: 'near bottom', with alternate floats (F) and 'poles' (float, float line (FL) and sinker (S)); and 'near surface' with alternate 'poles' and surface floats (SF). (Drawing by Stein Mortensen.)

line at intervals (e.g. 50 m), as shown in Fig. 3.10 ('near bottom'). The length of the sinkerline determines the depth of the mainline or in other words its height over the bottom. The skipper decides the length of sinkerline to be used in accordance with the vertical distribution of the fish. Another method includes the use of surface floats, often used when fishing relatively high in the water column (Fig. 3.10, 'near surface'). The surface floats with a line to the bight between two underwater floats give less depth variation of the mainline profile and have the same function as intermediate buoylines as hauling can be continued from the nearest surface float in case the mainline should break. Semipelagic longlining is used for species such as hake, pollock, cod and haddock.

Pelagic longlining

Pelagic longlining is the third main mode of operation. In contrast to bottom set and semipelagic longlines, pelagic longlines are normally not anchored but drift freely in the sea. Between the end (marker) buoys the mainline is suspended in the sea by floats and floatlines attached at intervals (Fig. 3.11). Pelagic longlines, called driftlines or floatlines, are mainly used in high-seas longline fisheries for pelagic species such as swordfish, tuna, shark and salmon, but are also used in coastal waters for species such as haddock during periods when the fish are feeding on pelagic prey.

Fig. 3.11 Pelagic longlining. (Drawing by Stein Mortensen.)

Other setting methods

The methods described above are the main ways of setting longline gear. There are numerous different local adaptations, of which a few examples will be briefly mentioned here to illustrate some variations from the main ways of setting longline gear.

Vertical longlining

This type of longlining is practised off Madeira in the fishery for black scabbard. The lines are set on steep slopes, with an anchor at one end and a float at the other end so that the mainline has a nearly vertical profile.

Monofilament longlining for pollock in South Korea

In this small-scale longline fishery, very thin monofilament lines are used. The lines are set semipelagically, with frequently attached surface floats. When hauling, all the surface floats are first picked up, and then the line is hauled as a number of line bights simultaneously. In this way the gear is brought safely and quickly on board. If the gear should be hauled the normal way (as one single line), the risk of breaking this very delicate line would be very high and the recovery of the gear could be very time consuming. This way of hauling naturally leads to a terrible tangling of the gear. Normally the gear is discarded after use and is thus probably one of the few examples of 'throw-away longlines'.

Overhaul longlines

This is a small-scale semipelagic monofilament longline fishery for cod in Newfoundland. Here the longlines are set permanently for a certain time period with daily overhauling. The lines are literally overhauled as the gear is hauled on one side of the vessel, the fish is taken off and the hooks are rebaited and then set over the other side of the vessel. One peculiarity in this fishery is also the use of hooks without traditional baits. Instead the line may be rigged with so-called 'feather hooks' – hooks with short tufts of nylon rope strands attached to the shank. This way of catching fish on longlines is naturally based not on attraction by smell stimuli but on visual feeding stimulation of fish that encounter the gear arbitrarily during migration or more local movements.

Safety lines

In the deep-water fisheries for hake, kingklip or Patagonian toothfish off southern Africa and southern America, longlines rigged with safety lines (also called top lines or mother lines) are used. The longline may be of monofilament or relatively thin multifilament. To avoid loss of gear the longline is suspended at intervals from a floated and much heavier (15–20 mm diameter) safety line. The gear is then hauled by the safety line. If the longline gets stuck in the bottom and breaks, it is retrieved at the next linkage point to the safety line.

One-end-anchor/buoy systems

This setting method is used as an alternative to the safety line system in the very deep water (1000–2500 m) longlining for Patagonian toothfish off Argentina. The lines are set down the steep slopes of canyons and grooves with a very rough bottom and a correspondingly high risk of line breakage. The anchor, buoy line and buoy are at the shallow end and only heavy weights (steel chains) are at the deep end of the longline fleet. In this way the heavy strain on the mainline caused by the last end anchor breaking loose is avoided. On the other hand there is no possibility of retrieving the gear from the far end after breakage of the mainline. The use of one anchor setting is therefore a question of evaluating the risk and safety aspects between the use of one or two end anchors. However, a clear benefit of the one-anchor system when fishing at great depths is the time saved as a result of not having to retrieve the last buoy line, which may be as long as 3000 m, since this means that more time can be spared for productive fishing (i.e. hauling of the longline instead of the buoyline).

3.3 The longline fishing cycle and modes of operation

General

The longline fishing cycle includes the following sequences, which are common for all longline operations: baiting, setting, fishing, hauling and maintenance.

Baiting

The hooks are baited either by hand or mechanically by baiting machines. Hand baiting is predominant and the baiting is commonly done on shore, but on larger vessels operating off shore the baiting is conducted on board. However, an increasing number of larger longline vessels are now using baiting machines.

Setting or shooting

The longlines are set from the stern of the vessel at speeds of 2–4 m/s (4–8 knots) according to the size of the vessel and the type of operation.

Fishing

The lines are left in the sea for between 1 and 24 hours (fishing or soak time) depending on the target species, the size of vessel and the number of hooks used.

Hauling (retrieval)

The lines are hauled, normally over the starboard side forward of the centre of the vessel. The hauling speed varies according to fishing depth, weather and current conditions, but the lines are normally hauled at speeds of 0.5–1.5 m/s (1–3 knots) with demersal longlining and 3–5 m/s (6–10 knots) with pelagic (tuna) longlining. During hauling, the fish are removed from the hooks, bled, dressed and iced or frozen.

Maintenance (repair or mending of the gear)

Before the lines can be rebaited, the old bait has to be removed, broken snoods and hooks have to be replaced and tangles of the mainline must be removed. At this point the mainline is also inspected for damage.

Baiting

Although the use of automatic baiting machines has increased in many longline fisheries (see Section 3.4, Mechanised longlining), baiting of longlines is still to a large extent done manually. Before fishing, a piece of bait has to be applied to each hook of the gear. As shown in Fig. 3.7, the baits may be either whole individuals of rather small bait organisms such as sardines, shrimps or mussels, or pieces of larger bait organisms, e.g. mackerel, saury or squid. When baiting the hooks it is very important to secure the bait as firmly as possible. When using bait organisms with tough flesh (e.g. mussels or squid), the method of applying the bait to the hook is not so critical as the hook usually gets a good grip of the bait independently of baiting technique. However, when using pieces of fish for bait (e.g. mackerel), care must be taken not only that the hook is fastened in the soft flesh, but also that the tougher skin part of the bait piece is penetrated once or twice by the point of the hook when applying the bait. If possible, the hook should also curve around the backbone of the piece of bait. Baits that do not stay well on the hook may reduce the fishing power of the longline significantly, either because they fall off during setting of the lines so that the number of effective hooks is reduced or because poorly secured baits are too easily removed from the hook by the fish so that they escape with the piece of bait rather than being caught.

When using larger bait organisms, cutting them into bait pieces of desired size is a significant part of the baiting process. With mackerel, a small fish (200 g) may be cut crosswise into, say, 10 pieces of about 20 g each whereas a larger fish is often split lengthwise before crosscutting, giving about 30 baits. Bait cutting is still often done manually, using a knife and a cutting board. However, bait cutting machines are widely used to regularise this work. The fish is dropped through a slot on to a series of parallel rotating cutting discs and is thus cut into equally wide bait pieces. Often such machines also have a device for splitting larger fish lengthwise before crosscutting. Nowadays bait is mostly available as frozen blocks (e.g. of herring, squid, etc.), and before cutting the bait it has to be thawed correctly. As a rule it should be thawed such that it is still a little on the icy side when it is cut, so that it does not soften too much before baiting.

Figure 3.12 shows a common situation from hand baiting of longlines: a woman rebaiting a skate of longline. In many fishing communities, onshore baiting of longlines provides an important source of employment. The unbaited line is overhauled, a bait is attached to every hook and the baited line is coiled into a tub or basket. For convenience a box with pre-cut bait is often hung on the edge of the tub for handy access to the bait pieces. Correct coiling of the mainline and

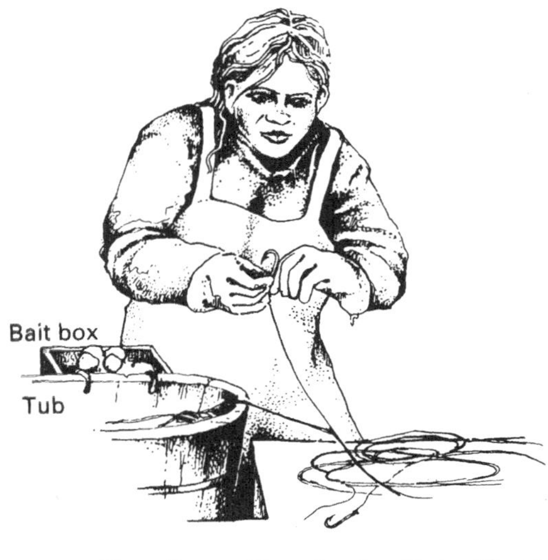

Fig. 3.12 Hand-baiting of longlines. (Drawing by Stein Mortensen.)

placement of the baited hooks in the tub is of great importance, and is a job that calls for skill if it is to be performed quickly and efficiently. If the job is not well done, there is a high risk of tangling during setting of the line. In the worst case, one of the hooks at the top of a baited skate may fall down through the coil of mainline. When setting the gear this hook may grab the whole coil of line so that a skate of several hundred hooks is set in one big tangle. The way in which the baited hooks are laid in the tub among the coils of mainline is therefore of great importance if tangling of the gear during setting is to be avoided. Anybody who has tried to bait longlines will have found that this is not an easy task for a beginner. When baiting thick multifilament longlines, it is common to lay the hooks along the rim of the tub on top of the expanding coils of mainline. It is important to cover the recently laid hooks with new coils of mainline as the baiting proceeds, in order to prevent tangling between hooks deep in the tub with parts of the mainline higher in the tub.

Baiting of monofilament lines and coiling them into tubs is a more difficult task. Because the monofilament lines are thin, rigid and spring back to their original shape after bending, the coils of line do not give the same support for the baited hooks as compared with thicker multifilament gear. The hooks therefore fall through more easily, and this may result in severe tangling. When baiting monofilament lines it is therefore common to lay paper sheets over the coils of baited gear for every 10–20 hooks during baiting, so that the hooks rest on the paper and are thus prevented from falling down through the underlying coils of line.

It may seem a little excessive to go into such detail about how baited hooks are laid in the tubs. However, the catching efficiency of longlines is a result of careful preparation of different details of the gear, including coiling of baited lines. If this is carelessly done, the gear may be set with numerous tangles of a few to more than a hundred hooks, which significantly reduces the catching power of the gear.

When hand baiting into tubs is done ashore, the total number of skates or tubs needed for a fishing trip is baited before the vessel goes to sea. Depending on vessel size, the amount of gear may vary from a few to more than a hundred tubs of longline. The time for baiting may therefore be hours and even days depending on the number of people (baiters) engaged in the work. The baited lines are therefore usually kept refrigerated or frozen until the vessel is ready to sail.

Tubs are the traditional and still widely used solution for storage and transportation of the gear in many longline fisheries in different countries. A classical tub design is shown in Fig. 3.13. This type of tub is traditionally used in the Norwegian and north west Atlantic longline fisheries. The shape and size of longline tubs may differ slightly between regions. Tubs used in eastern Canada are of the same basic shape, but are taller and more slender. Japanese tubs are also round, but

Fig. 3.13 Wooden longline tub.

are usually smaller (40 cm diameter, 20 cm height). In Portugal and Spain it is not unusual to find relatively small square-shaped wooden tubs (or baskets), with a 30 cm square base and a 40 cm square top. There are usually a few holes in the bottom of the tub for drainage. Wooden tubs are often painted to make cleaning easier. In recent years tubs made of different plastic materials have come to be used more frequently as an alternative to wooden tubs. However, many fishermen still prefer the wooden tubs as they have properties that are not easily matched by the plastic versions. The higher weight of wooden tubs and the nature of the material make them more steady on the deck of the vessel so there is less sliding about. Furthermore, in the long run, wooden tubs have also proven to be more durable. However, improved plastics are being developed, and plastic tubs will most probably predominate in the future, provided that longlining with the use of tubs survives the competition from mechanised longline systems.

Baiting and coiling of the lines into tubs is mainly used with longlines where the snoods and hooks are fixed permanently to the mainline and/or longlines with a relatively short hook spacing. Longlines with wider hook spacing, either with fixed or 'snap-on' gear (the snood and hook being detached during hauling and attached during setting) are often baited during setting of the gear. Typical examples are the oceanic longlining for tuna and the longline fishery for Pacific halibut, where the snoods with baited hooks are manually attached (or 'snapped on') the mainline during setting. Also longlines with fixed snoods but with a fairly wide hook spacing, enable baiting of the hooks as the gear is set. The hooks are then fastened in the correct sequence along the rim of the tub or basket, and the wide hook spacing leaves enough time between each hook to be baited during setting – even at fairly high vessel speeds. To make this work a little less stressful and dangerous, it is often shared by two fishermen who each bait and let go alternate hooks.

Setting

Setting or shooting the gear is normally done from the stern of the longline vessel, regardless of vessel size and type. Time and place of setting the gear is based on experience, and in recent years the decision has also been aided by modern technology such as electronic fish-finding equipment (see Section 3.6). With demersal longlining, the fish is in most cases not distinguishable from the bottom when using echo sounders, so the main use of echo sounding in longline fishing is still to find the correct depth for setting the gear. Satellite navigation equipment also helps the longline skipper to position the gear precisely according to earlier catch records. Similarly satellite maps of ocean surface temperatures are used for selecting fishing areas in the longline fisheries for tuna.

Time of setting is also of importance in longline fishing. The feeding motivation of fish varies with time of day, and so do the feeding periods of bait scavengers. In the longline fishery for tusk and ling, for instance, the skippers try to set most of the gear shortly before dawn, which corresponds to the main feeding period of the target species and avoids the feeding period of most scavengers, which are mainly night feeders.

When ready for setting, the end marker buoy and floats are dropped as the vessel moves slowly forward. Then the buoy line and, for demersal or semipelagic longlining, the anchor are dropped, before the vessel speeds up and the setting of the gear starts, normally at a speed of 4–10 knots (2–5 m/s). The lines are shot along a predestined course track, which may be changed during setting if the skipper observes unexpected depths or bottom conditions with the echo sounder. When there are a few hooks of line left to shoot, the vessel speed is usually reduced before the end anchor, buoy line and marker buoy are dropped. Although setting of the gear is the least time consuming part of the longline fishing cycle, the setting of the daily gear amount takes 3–4 hours with large scale demersal longliners or in the tuna longlining, where in both cases more than 50 km of line is fished daily. In general the setting operation is fairly similar for most longline fisheries. However, there are operational differences between the setting of prebaited lines (tub gear), manual baiting during setting, and mechanised (automatic) baiting. The latter case is described in Section 3.4. Semipelagic longlining for cod and tuna is described here as examples to illustrate the setting of prebaited lines, and baiting during setting, respectively.

Setting semipelagic longlines for cod

As soon as the anchor is dropped, the prebaited line pays out at vessel speed and if the baiting job is done properly the line is set without severe tangles. A fleet consists of several skates separately baited and coiled into tubs. During setting these have to be linked. To be able to do this in a practical way during setting, a piece (1 m) of the bottom end of the line is hung over the rim of the tub when baiting. When the baiting is completed, the bottom and top ends of the skate are tied together and laid on top of the line coil in the tub. The ends of the skate are thus easily identified when preparing the skate for setting. The top end of the first tub is tied to a short rope linked to the first anchor; the bottom end is tied to the top end of tub number two and so on until the fleet is completed and the bottom end of the last tub is tied to the last anchor. When setting the lines, it is critical that the hooks run smoothly overboard without catching on the railing, constructions on deck or people. With setting of tub gear, the so-called line setters or line chutes are used to safeguard the lines from the tubs over the stern of the vessel. The principle of the line chute is shown in Fig. 3.14. In earlier times the chutes were made of wood with an inner lining of sheet metal to provide a hard and smooth surface that did not allow the hooks to penetrate and get stuck. Today most line chutes are made from aluminium or steel. Line chutes can be single or double (Fig. 3.14, bottom). When using single chutes, the tubs are lined up in front of the chute in one line, and in two lines with double chutes. The double chute provides better safety for the fisherman as he does not have to remove the empty tub in front of the hooks paying out from the next tub in line.

The floats and sinkers used in semipelagic longlining are attached at intervals for each skate during setting. This is made possible by short loops of rope attached to the mainline at the desired intervals. When baiting, the ends of these loops are fixed in the right order along the rim of the tub. This arrangement gives the fisherman the time needed (a matter of seconds) to attach the float/sinker lines which are thrown over the stern as the attachment point on the mainline leaves the tub. With demersal longlining, small (1–2 kg) weights are often attached to the mainline as well as intermediate anchors with buoy lines and buoys. Here the intermediate anchors and the weights are attached at the link between skates, the weights often between every skate and intermediate buoylines at every five to ten skates.

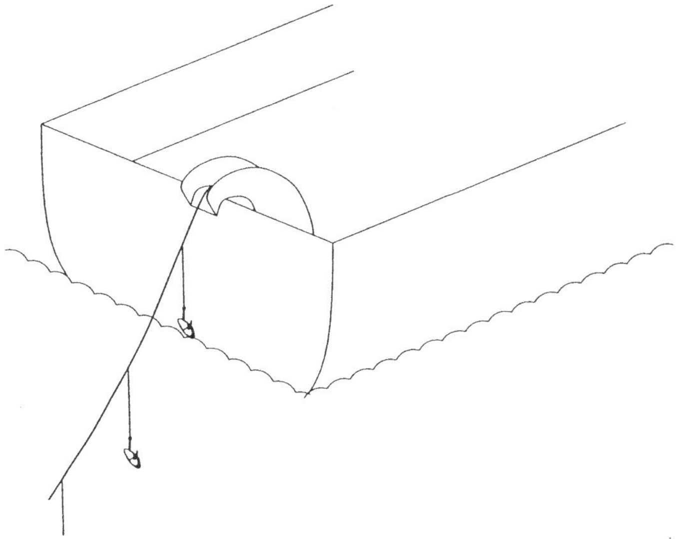

Fig. 3.14 Longline chute and tubs arranged for setting.

Setting semipelagic longlines for tuna

With wide hook spacings, as in the longline fishery for Pacific halibut or in the tuna fishery, there is time available to bait the hooks during setting. The use of detachable snoods or so-called 'snap-on gear' (Fig. 3.1(c)) is common with this type of longlining. When setting, the snoods with baited hooks are snapped on to the mainline as this pays out at vessel speed. Even with very wide hook spacings it may be very hard for one man to keep up the high pace of baiting and snapping on every hook. This work is therefore commonly shared by two or three fishermen who alternately snap on every two or three snoods with baited hooks. To secure a relatively even hook spacing, a timer with a sound signal tells when to attach the gangions. On the large tuna longliners, a motorised hauler (line shooter) is also commonly used for paying out the mainline over the stern of the vessel. The hauler speed is adjusted so that the line pays out somewhat faster than the actual vessel speed. This is done to obtain a constant shooting speed of the mainline during setting, as the vessel speed may vary considerably with the pitch of the vessel – particularly when setting under rough sea conditions.

When setting longlines, regardless of setting method, the baited hooks are available targets for sea birds for a few seconds before they sink far enough to be out of reach. Depending on the sinking speed of the actual line and the size of the vessel (the higher above sea level the line leaves the vessel, the further behind the vessel the line is allowed to sink), the baited hooks may be available for sea birds over a distance of between 50 and 100 m. To avoid the baits being removed by sea birds and to prevent birds from accidentally being caught on the line, a variety of methods can be used to scare birds away from the line. A common solution is to trail a float or pieces of clothing parallel to the longline at the actual distance behind the vessel (see also Section 5.8).

Hauling

Hauling or retrieving the gear, which is by far the most time-consuming part of longline fishing, starts by locating the marker buoy. Aided by modern navigation instruments the buoys are usually easily recovered, even in foggy and dismal weather. In darkness, powerful searchlights are also used to spot the buoys effectively in the near field. The marker buoy is picked up, either directly with smaller vessels, or from larger vessels by using a throwing hook on a line or a long pole with a treble hook to get hold of the buoy line. Then the buoy line is hauled and the first end anchor is retrieved before the hauling of the longline can start. In small-scale and shallow-water longlining, the hauling may still be done by hand. However, in most longline fisheries today, the hauling is done by so-called line haulers: powered capstans with specially designed sheaves for holding and pulling in the rope or line. Figure 3.15 shows a common deck arrangement at the hauling position of a longline vessel, with the rail roller and some typical hauler designs with details of the line sheave. The function of the rail roller, which consists of one (main) horizontal roller and two vertical side rollers, is mainly to protect the railing of the vessel and the line from chafing, and to ease the pull by eliminating the 'railing to rope friction'. The side rollers keep the line from leaving the main roller. The fore–aft position of the line on the main roller is also a good indicator for the skipper of the position of the vessel relative to the line. If the line comes in over the front part of the main roller, the vessel speed is too low relative to the hauling speed. Ideally the line should come in over the aft part of the main roller, close to the aft side roller.

From the rail roller the line goes to the hauler. The line is squeezed into the wedge-shaped sheave, so that the sheave gets a good grip of the line. If the wedge of the sheave is correctly designed, the grip of the line gets harder with increasing force of pull. However, as the rope travels around with the rotation of the sheave, the grip of the line is loosened, but normally not enough to let the line loose before it has travelled 360° with the sheave. The line must therefore be released from the sheave, either manually by a fisherman who pulls the line from the hauler and coils it into a tub, or by automatic coiling devices. With horizontal sheaves a so-called 'slack hauler' is often used, with two rotating wheels that grip the line, pull it from the main hauler and coil it into the tub. With vertical sheaves the line coils directly into the tub after being released from the hauler by a stationary metal wedge that fits into the groove of the sheave. In the first part of this century the line haulers were mechanically driven with transmission systems from the main engine. Today the haulers are electrically or, most commonly, hydraulically powered.

The de-hooker and the hook cleaner are also common devices in the longline hauling equipment, and are positioned between the rail roller and the hauler. The de-hooker consists of two narrowly spaced vertical bars or rollers that allow the line and hooks to pass, while fish that are not released from the line by the fisherman will be torn off the hook at this point. For removal of remaining baits, hook cleaners are often used in addition to the de-hooking device. The most commonly used solution is a set of closely spaced, facing, stiff brushes, where most of the old bait is cleaned off as the longline is hauled through.

The manual labour left in the hauling process is the de-hooking and bleeding of fish and some other tasks done by the fisherman working at the hauling position. His main tools are the gaff hook (Fig. 3.16) and a sharp knife. Smaller fish may pass over the rail roller and be released from the line by the de-hooking device. Larger fish are, however, normally de-hooked manually at the rail roller by means of the gaff hook. As the fish leaves the water, the increased weight strain often leads to loss of fish, particularly as the fish hits the rail roller. De-hooking of the fish outside the railing is therefore important to avoid loss of fish or the extra work of picking it up from the sea surface. However, it is not possible to avoid some fish falling off and these, especially species that have a swimbladder, will then float at the side of the vessel. These fish can usually be saved by the use of the 'long-pole' (a 2–4 m pole of bamboo, wood or fibreglass with a treble hook at one end).

The main work of the fisherman at the rail roller is therefore to gaff the fish, bring it on board, bleed it by cutting the throat and drop it in the bleeding bin. In addition he must clear tangles as these do not pass easily through the hauling system; adjust the speed of the hauler; replace the full tub with an empty one when a skate is hauled; and otherwise ensure that the hauling runs smoothly.

With semipelagic and pelagic longlining, the intermediate buoys and buoy/sinker lines have to be detached and handed over to another crewmember for hauling these in, either by hand or by a separate line hauler. With the use of 'snap-on gear' as in tuna longlining, the snoods are detached and hauled in by a separate snood hauler. The hauling of a longline fleet is completed by retrieving the end anchor, buoy line and end buoy with additional floats before the vessel steams to the next fleet of gear to be hauled or set.

Between hauling and setting, the gear must be maintained. With tub longlining the maintenance is done as the lines are rebaited. When overhauling and rebaiting the line, every unit piece of the gear is inspected. If the mainline, snood and hook look fine, a new bait is

(a)

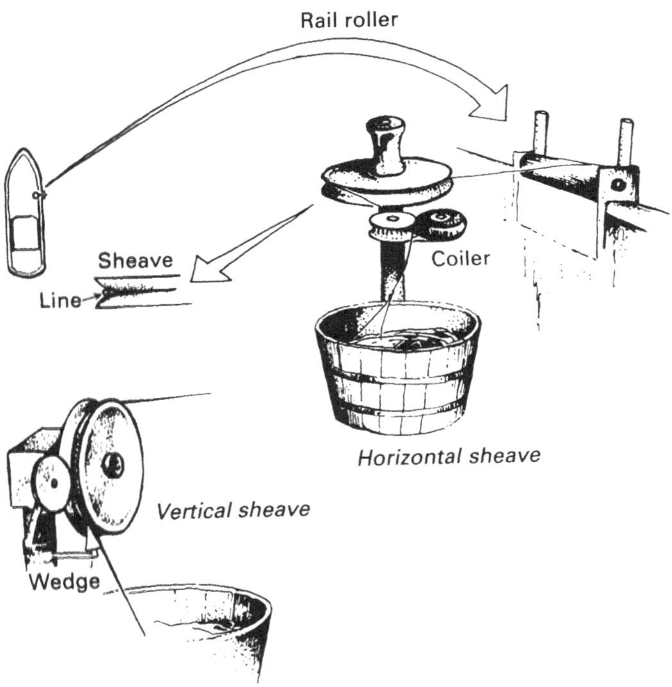

Fig. 3.15 Longline hauling equipment. (a) Basic haulers (vertical and horizontal) and rail roller (drawing by Stein Mortensen); (b) basic gear configuration and operational equipment in the longline fishery for Pacific halibut. (Drawing by C.H. Hitz, from Myhre and Quinn, 1984.)

(b)

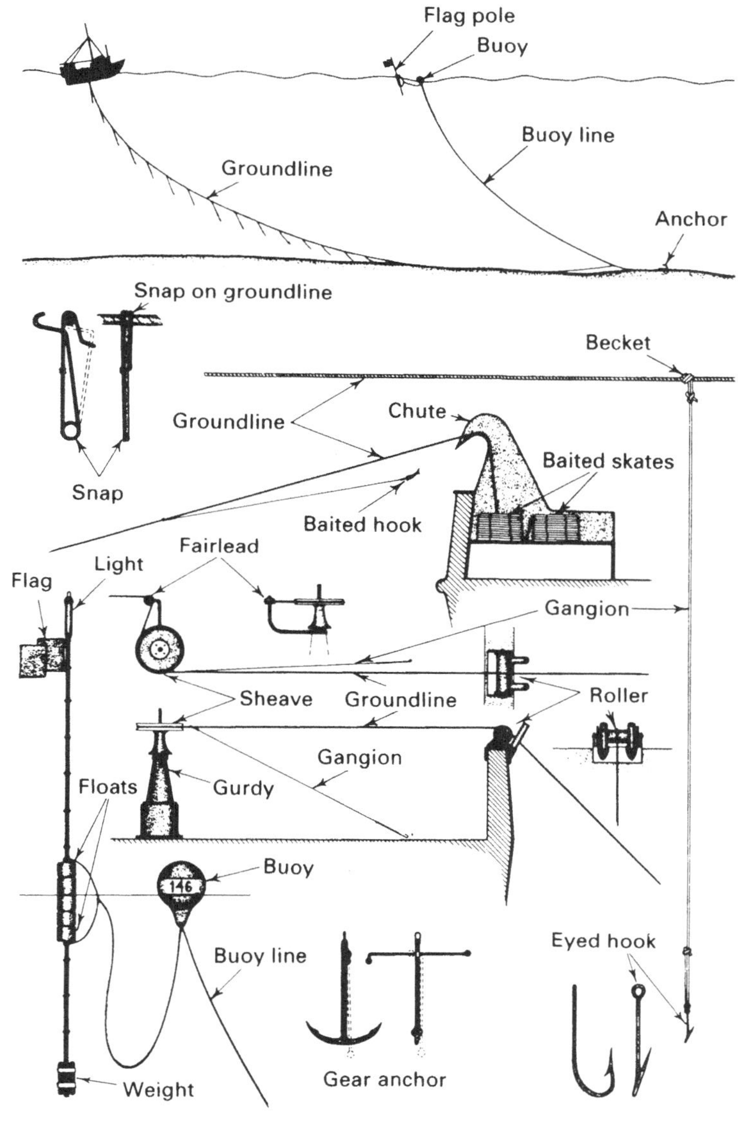

placed on the hook after removal of remnants of old bait. The types of damage that occur most frequently are broken or misshapen hooks and broken snoods. If a hook is not broken or not too much damaged, it is normally reshaped. This is done either by bending it over a fixed nail or by using a specially designed tool – a so-called hook re-shaper (Fig. 3.17). This mending tool is designed for the specific hook type being used, so that the original shape is regained. When the hook and/or snood needs to be replaced, a complete new snood and hook unit is mounted. In any longline operation an ample supply of pre-rigged

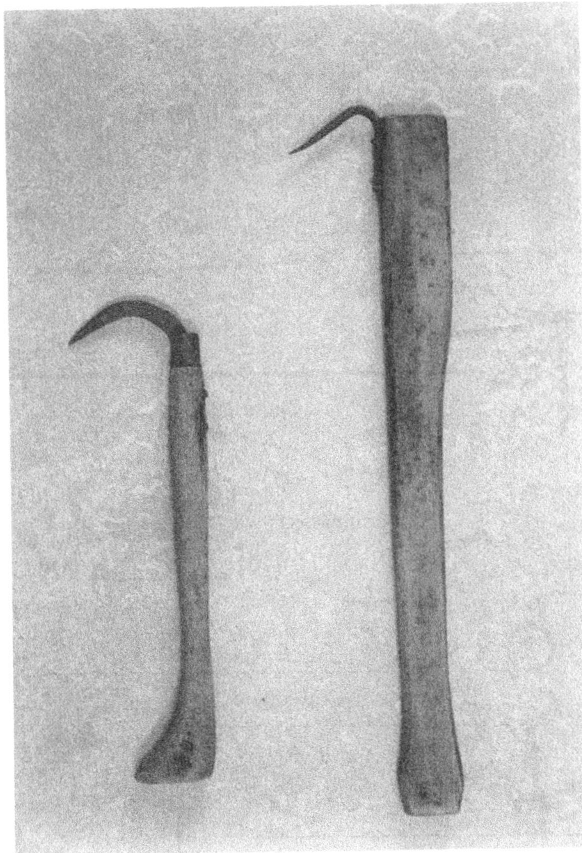

Fig. 3.16 Gaff hooks: Korean (left) and Norwegian (right) types.

snood and hook units is needed. If the mainline is damaged (for example it is heavily chafed or one or two strands are broken), the damaged piece is cut away and the line ends are spliced together.

When hauling, careful manoeuvring of the vessel is an important success factor in longline fishing. The course of the vessel should be parallel or slightly to the left of the direction of the longline fleet being hauled, and the vessel speed should be adjusted to the hauling speed (or vice versa) so that the vertical profile of the line has a slightly backwards curve relative to the vessel direction. Examples of favourable and unfavourable vessel-to-line situations are shown in Fig. 3.18. The 'correct' situation is indicated by the vessel-to-line angle A2 and vertical profile B2. With this situation, the skipper has good control of the vessel-to-line relationship. By observing that the line comes in over the aft part of the rail roller and that it points slightly backwards and slightly out from the side of the vessel, he knows that the line is in an

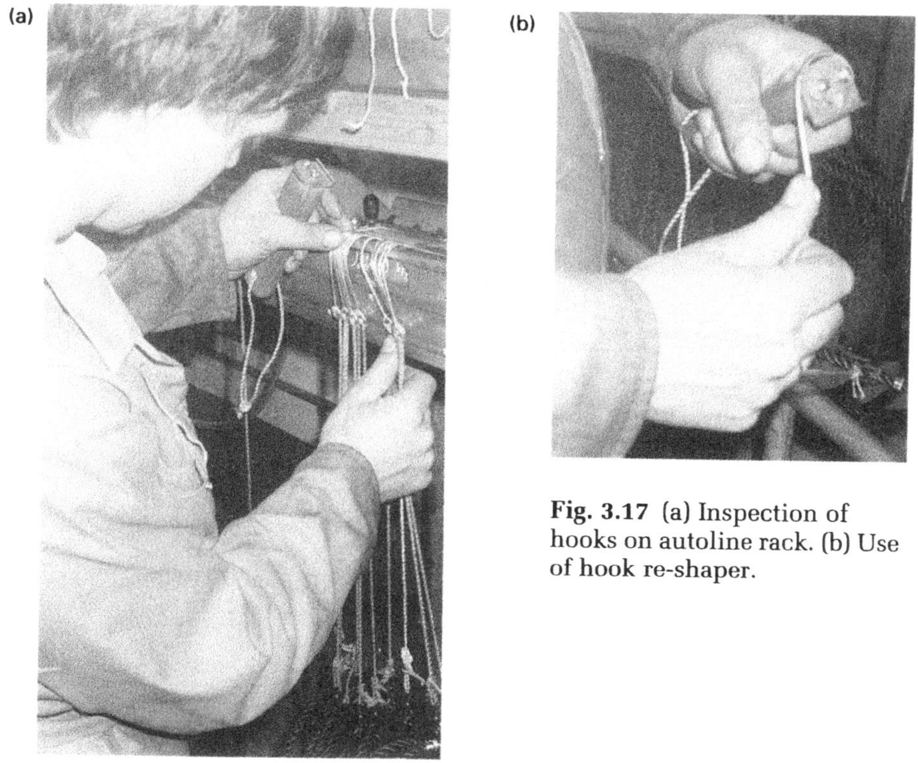

Fig. 3.17 (a) Inspection of hooks on autoline rack. (b) Use of hook re-shaper.

A2–B2 position. The curve of the B2 profile also provides sufficient elasticity (slack) to reduce the risk of breaking the mainline with the pitching and rolling of the vessel.

When hauling the line over the starboard side of the vessel, vessel-to-line angles to the port (left) side of the vessel (A4, Fig. 3.18) are always unfavourable because the line is pulled against the hull of the vessel. In this situation the skipper has no indication of the angle and there is a high risk of losing fish as they may be knocked off when hitting the keel or other parts of the hull. The B1 situation increases the risk of the line being caught by the propeller and of dragging the line for a long distance at the sea surface, which can easily lead to loss of fish. In the B3 and A3 situations, the flexibility that is provided by the (B1,B2) curved and slack line profiles is lost, with increased risk of breaking the mainline if, for instance, it became stuck on the sea bed.

Smaller longline vessels are usually manoeuvred from the working deck during hauling, often by the person at the rail roller. On larger longliners, the handling of the vessel is done from the bridge by the skipper or the mate.

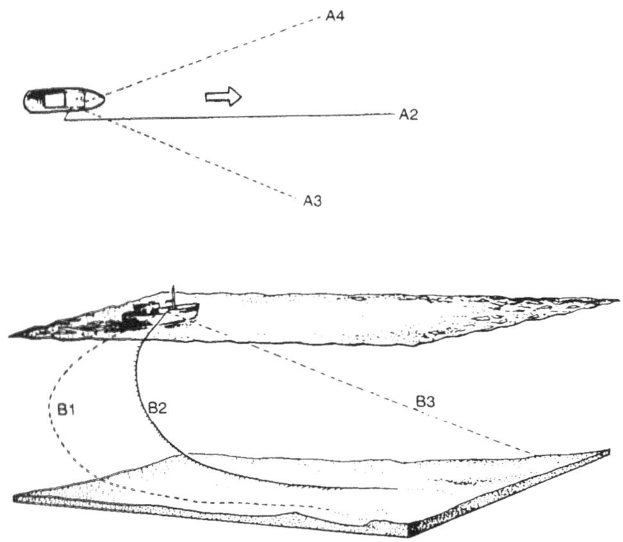

Fig. 3.18 Vessel-to-line situations during hauling. (Drawing by Stein Mortensen.)

3.4 *Mechanised longlining*

The term mechanised or automated longlining is used to describe technical systems that rationalise the manual labour of longline fishing beyond the use of the equipment described above (i.e. hydraulic haulers and line coiling devices). The term 'mechanised longlining' most commonly refers to systems or machines for automated baiting. However, even with the most advanced mechanised systems at present, a major part of the longline operation is still based on manual labour, the main advantage of the mechanised systems being improved effectiveness and simplified handling of the gear, without necessarily reducing the number of crewmembers.

In the efforts to mechanise the tedious work of baiting longlines, two main problems had to be overcome:

- baiting: to design and construct systems for applying a piece of bait to the hook;
- gear handling and storage: to design and construct systems for handling and storage of the line between the hauling and baiting process in such a way that the gear could be run effectively from the storage system to the baiting unit.

Since the early 1960s substantial effort has been put into research and development work on a large number of different mechanised longline systems in Canada, the Faroe Islands, Germany, Iceland, Ireland, Japan,

Norway, Sweden, the UK and the USA. For different reasons most of these systems were never brought beyond the development stage, and only a few are used in commercial longline fishing at present. However, when reviewing the different solutions, they are all based on one of two main principles of automated baiting (precise or random baiting) and on one of two main principles for storage of the gear (rack or drum storage). These main principles are shown in Fig. 3.19.

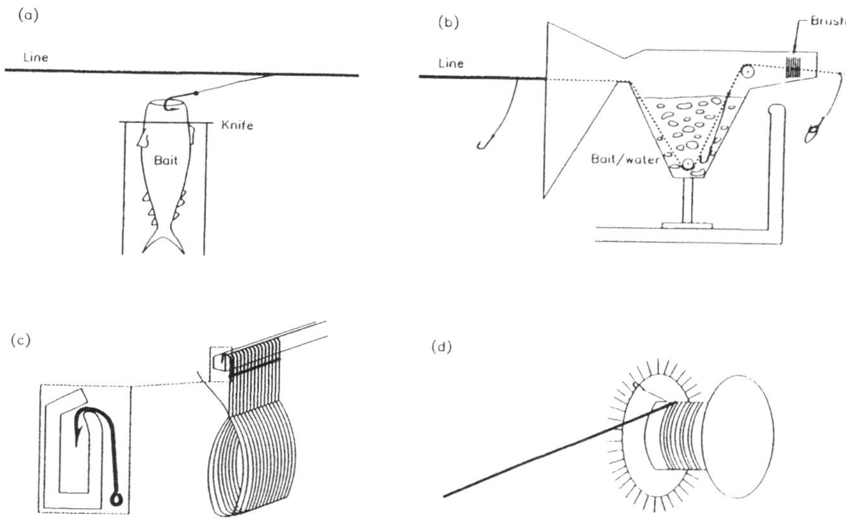

Fig. 3.19 Main principles of mechanised baiting and gear storage: (a) precise baiting; (b) random baiting; (c) rack storage; (d) drum storage.

With precise baiting, a piece of bait is applied to each individual hook by a standardised mechanical operation. The baiting quality should be comparable to that of qualified hand baiting. With random baiting, the longline is set through a specially designed container with a mixture of pre-cut bait and water. As the hooks pass through the bait mixture, they snag a piece of bait. The term 'random' refers to the non-controlled nature of this process. The hooks snag the pieces of bait at random, which leads to variation in baiting quality – from baits being securely fastened, to baits that barely stay on the hooks.

Rack storage is based on the use of metal racks or rails which are profiled to the hook type being used. The hooks are put on to the rack in sequence, with the snoods and loops of mainline suspended underneath. Drums are used either to store the mainline only (with the use of detachable snoods) or to store the complete gear with some arrangement for storage of hooks on the drum.

A total of more than 40 different mechanised longline systems have been developed. It is beyond the scope of this book to describe them all.

Furthermore there are great similarities between many of the systems. Some systems, with special characteristics within the different combinations of baiting and storage principles, are therefore chosen as examples to illustrate various solutions to mechanised longlining.

Systems based on precise baiting and rack storage

The first automated longline system to be used in commercial fishing on a regular basis was the Autoline system manufactured by O. Mustad & Son Ltd of Norway. This is still the most widely used and distributed mechanised longline system and will therefore be described in more detail to illustrate the design and function of systems based on rack storage and precise baiting. The principles of the Autoline system are shown in Fig. 3.20. When setting, the line is pulled from the storage racks through the baiting machine and over the stern of the vessel by the drag of the anchor and buoyline. Bait fish such as whole mackerel is laid manually on a conveyor belt that feeds it stepwise into the baiting unit. The feeding steps may be regulated according to the desired width of the individual baits. Figure 3.20 shows how the bait is held with a spring-loaded 'finger'. As the hook penetrates the bait, a piece of bait is cut by a mechanically driven knife and the baited hook is pulled out of the machine. This operation can be run at a speed of 3–4 hooks per second. (A setting speed of 8 knots corresponds to 3 hooks per second with a hook spacing of 1.3 m.) When setting, there are normally two people working at the baiting machine, one for feeding bait fish on to the conveyor belt and one person to check that the hooks run smoothly off the rack, clearing potential tangles and replacing empty racks with loaded ones as the setting proceeds. A third crewmember supplies adequate amounts of bait and organises the anchors, buoy lines and buoys. Before the setting starts, the required amount of bait has to be thawed to the right consistency (around the melting point so that the bait is still rigid but not too soft).

When hauling, the work at the rail roller is no different from traditional longlining, with one fisherman gaffing and bleeding the fish. From the rail roller the line passes the de-hooker and the hook cleaner. It then goes via the hauler to the twist remover and the hook separator unit. When the line is retrieved, most of the snoods are twisted around the mainline. In traditional longlining the twisted snoods are cleared from the mainline by hand. This process is done automatically by the twist remover of the Autoline system, using water jets flushing the line at an angle towards the hauling direction. Cleaning the line is obtained as a side effect of the twist removal. The hook separator catches most of the hooks and guides them on to rack magazines.

When passing over the sheave of the hook separator, most of the hooks are automatically caught and guided on to the storage rack. This is done by the use of magnets that attract the hooks and lead them on to a guiding rail from which they are gravity-fed on to the rack. As shown

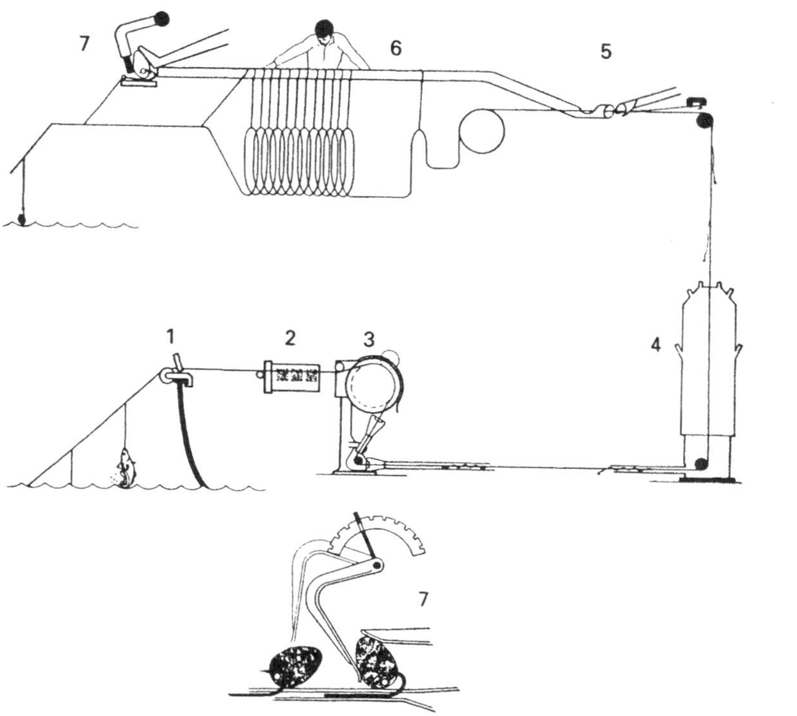

Fig. 3.20 The principles of the Mustad Autoline system: (1) rail roller; (2) dehooker and hook cleaner; (3) hauler; (4) twist remover; (5) hook separator; (6) storage rack; (7) baiting machine (expanded view: spring-loaded 'finger' keeping the bait in position for the approaching hook).

in Fig. 3.20, rollers and pipes with end funnels are used for safe guidance of the longline from the hauling position to the storage racks.

When the hooks and line arrive on the storage rack, the repair work starts. Three crewmembers are constantly occupied with repair work during hauling, inspecting hooks, snoods and line and conducting the necessary repair work as described previously for traditional longlining. The repair work is therefore the most laborious part of longline fishing with an autoline system.

Several systems have been developed on the Autoline model. However, only a few of these have been used commercially. To the authors' knowledge, the only alternative systems of this kind being used in commercial longlining at present are the Marco-matic and Circle-matic systems manufactured by the Marco Company (USA).

Systems based on precise baiting and drum storage of the line

There are presently no automated longline systems based on this combination in commercial use. However, to illustrate solutions using this combination of principles, four different developments are briefly described: the Autoclip, the Turboline, the Isifisk and the Mustad Miniline systems.

The Autoclip system

The Autoclip system (Fig. 3.21) was developed by the White Fish Authority, UK. It was based on detachable snoods, using a nylon clip that was automatically squeezed on to the mainline during setting of the gear. When the gear was hauled, the snoods were automatically detached from the mainline and stored in cassettes, while the naked mainline was reeled on to a drum. When setting, the cassettes were attached to the baiting unit and the snoods and hooks were attached to the mainline at intervals and run through the precise baiting machine. An advantage of the Autoclip system was the option of variable hook spacing that could be adjusted according to the type of fishery and the fish density. However, the system was not brought to a stage where it was used in regular commercial longlining.

The Turboline system

The Turboline system, which was developed by Bjørshol International (Norway), is based on detachable hooks (Fig. 3.22). The main element of

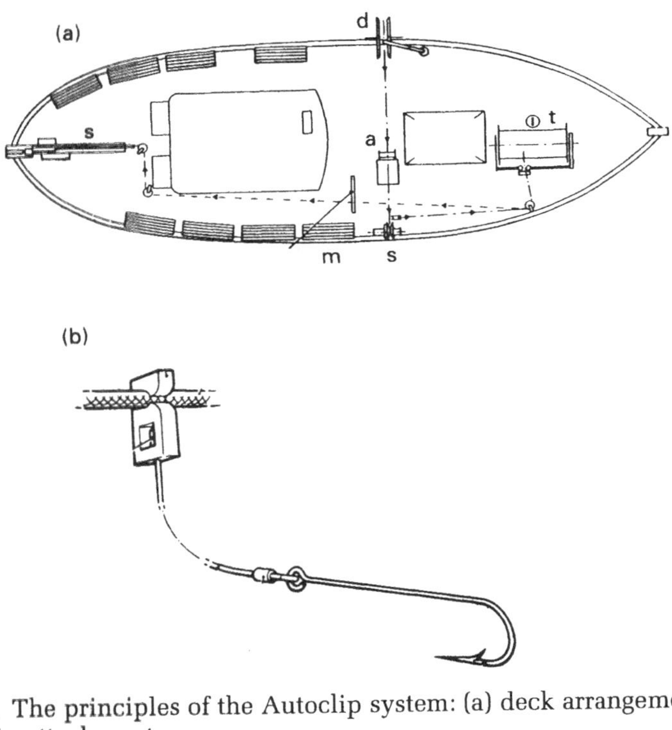

Fig. 3.21 The principles of the Autoclip system: (a) deck arrangement; (b) nylon clip attachment.

(a)

(b)

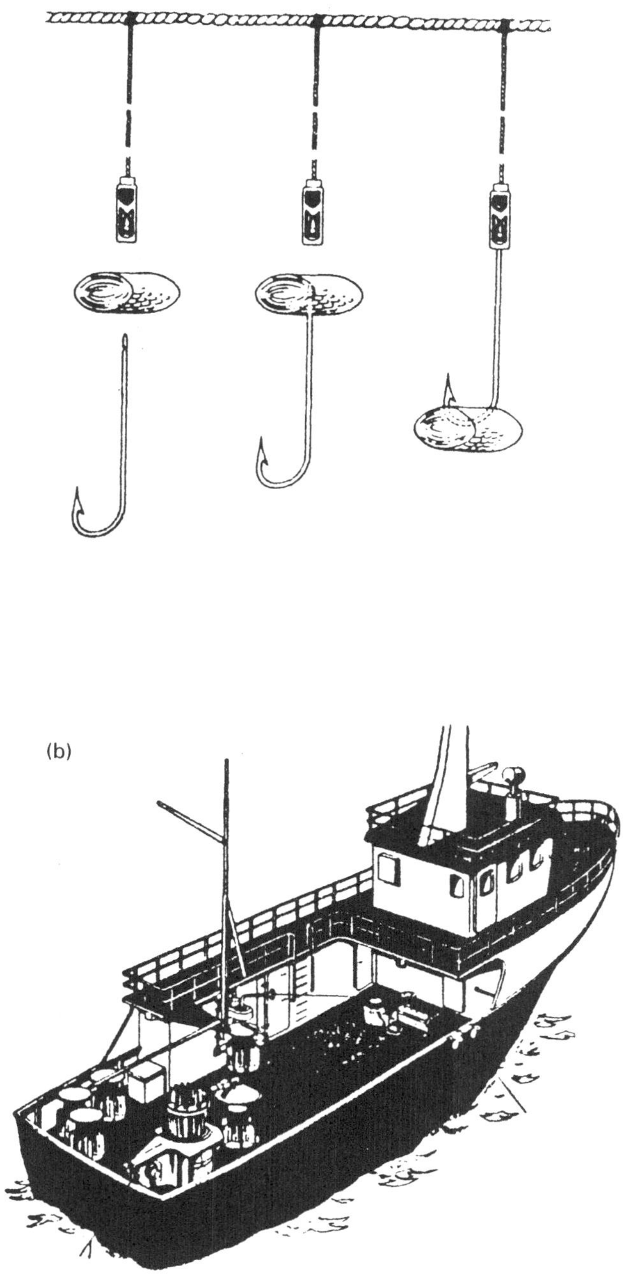

Fig. 3.22 The principles of the Turboline system: (a) hook coupling principle; (b) deck arrangement.

the Turboline system is the coupling unit that enables automatic detachment and attachment of hooks with a special arrowhead termination of the shank. When hauling, the hooks are detached and stored in special magazines, while the mainline and snoods are spooled on to a reel. When setting, the hooks are shot backwards through the bait into the coupling unit. To the authors' knowledge the Turboline system is still under development, and has not yet been applied in regular commercial longline operations.

The Isifisk system

The Isifisk system is also still at the development and trial stage. It is being developed by the inventor, Mr J.J. Dahl, and the Johnsen Bros Company (Norway). The system is primarily for automated operation of monofilament longlines. As shown in Fig. 3.23, it is based on drum storage of the complete gear. During hauling, the mainline is spooled on to the drum while the hooks are guided on to radial racks that also act as the lower shield of the drum. The system also includes an automatic repair unit for replacement of broken snoods and hooks. When setting, the line is pulled off the drum and through an automatic baiting unit. The system shows promise but there are still problems to be solved before it can be regarded as sufficiently reliable for use in commercial fishing.

The Mustad Miniline system

The Mustad Miniline system was originally designed in the late 1970s for handling and baiting monofilament gear. The snoods and hooks are

Fig. 3.23 The Isifisk system.

detached during hauling and the line is spooled on to a drum. The hooks are then baited by hand and stored in cassettes. When setting the gear, the line is pulled off the drum and as the specially designed attachment devices pass through a setting apparatus with a cassette connected to it, a snood with a prebaited hook is attached to the line. This concept has been abandoned and the Miniline has been tried with random baiting, without gaining any popularity in commercial long-lining so far. However, a totally new version of the Miniline concept with precise baiting has recently been developed. The principle of the system is shown in Fig. 3.24. As with the original design, it is based on snoods that are detachable by means of special coupling devices at the snood terminations that are attached/detached to and from corresponding coupling units which are mounted at intervals (1.8 m) along the mainline. When hauling, the process is similar to other longline operations until the line has passed the hook cleaner. Then the snood and hook are automatically detached from the mainline and are guided on to a double rack, one rack for the hooks and one for the coupling device, while the naked mainline is reeled on to a drum. When setting, the snood and hook racks are linked to the precise baiting machine. As the mainline pays out from the drum, a hook and snood unit is automatically attached to the special attachment device and baited as it

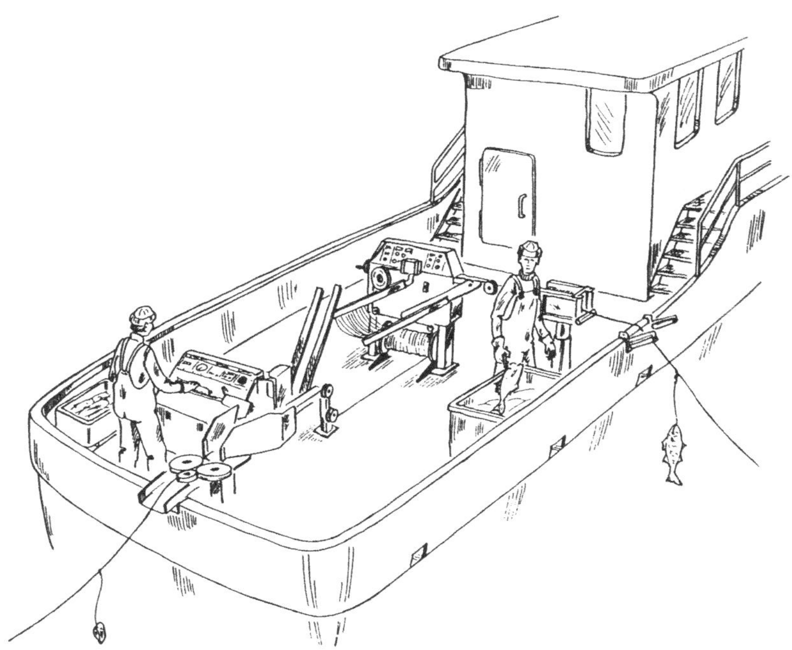

Fig. 3.24 The principles of the Mustad Miniline system. (Drawing by Helge Iversen.)

passes the baiting unit. The hook spacing can be increased stepwise by a function on the baiting unit where the fisherman can choose to have hooks on every second, third or fourth attachment device, thereby increasing the hook spacing from the regular 1.8 m to 3.6 m, 5.4 m, 7.2 m and even wider spacings.

Systems based on random baiting and rack storage of the line

Different random baiter systems became popular in the eastern Canadian longline fisheries during the early 1980s. These were relatively inexpensive systems as compared with the cost of precise baiting machines, and were reasonable alternatives for smaller vessels (9–18 m; 30–60 ft). The random baiter was also introduced in other countries, e.g. Norway, as an alternative to hand baiting in small-scale coastal longlining. Different designs of random baiters were manufactured and mainly used with rack storage of the line. A random baiter is shown in Fig. 3.19(b). This and other random baiting systems could be used either with the common rack storage arrangement or with racks vertically mounted on the inside of the tub, so that the lines could be set from tubs through the random baiter. The random baiter is filled with pre-cut bait, usually mixed with water. To prevent loose pieces of bait from being dragged out by the line and baited hooks, most random baiters are equipped with brushes at the aft end. The brushes also have the function of pushing loosely fastened baits more firmly on to the hook.

Systems based on random baiting and drum storage of the line

There are few if any systems developed with this baiting/storage combination. However, most of the different drum systems described above could be used with a random baiter unit.

Success factors of mechanised longlining

The use of mechanised longline systems has steadily increased since the first successful trials with the Mustad Autoline system in the early 1970s. However, automated systems are still restricted to a relatively small fraction of the world's longline vessels that have certain characteristics in common. Thus they are mainly found in vessels that operate longlines as a full-time or close to full-time occupation, using multifilament gear with a narrow hook spacing (1.2–1.4 m). The first Autoline systems were successfully used in the longline fishery for spiny dogfish in the North Sea. These were vessels that also fished for tusk and ling. Gradually most of the Norwegian fleet of larger longliners were equipped with Autoline systems, followed by Faroese, Icelandic and a few French and British vessels, with cod, haddock, tusk and ling as the most important target species. Today this fleet of larger autoliners fishing the North Sea and north east Atlantic comprises nearly 100 vessels. Since the late 1980s there has also been a growing fleet of mechanised longline vessels or factory longliners in the northern Pacific. These are mainly US-registered vessels fishing for Pacific cod in Alaskan waters, but there are also Russian and Japanese vessels fishing in the Bering Sea or the Sea of Okhotsk. In addition, large mechanised longliners are found in the hake fishery off Chile, in the fishery for Patagonian toothfish off Argentina and in the longline fishery for kingclip off New Zealand. Fully automated or mechanised longlining is still restricted to a relatively small number of vessels because of the cost or because of the restricted versatility of gear that can be used with today's mechanised systems. Mechanised longline vessels operating in the Atlantic are predominantly equipped with the Mustad Autoline system, while in the Pacific both the Mustad and the Marco systems are used.

As stated previously, mechanised systems are mainly found on larger vessels [30–50 m length overall (LOA)]. This is not because space restrictions do not allow an autoline system to be installed on smaller vessels, as the systems are being used successfully on longliners down to 15 m LOA with three or four crewmembers. However, the smaller the vessel, the higher the relative investment in mechanisation equipment as compared with the cost of the vessel and the yearly income. To illustrate this, the new building costs of larger mechanised vessels are in the range US$5–15 million, with corresponding yearly earnings ranging from US$2 million to 15 million. The cost of a fully mechanised system with precise baiting is about US$150 000–200 000, which gives a relative cost of 1–4% and 1–10% of the total vessel cost and annual income, respectively. With a smaller vessel (15 m LOA) with a new building price of US$0.5–1 million, the relative cost of a fully mechanised system is 15–40% of the vessel cost. As proven by some smaller vessels using the Autoline system, investments of this relative size can be justified. However, this requires that the vessel be occupied full time with mechanised longlining so that the system can be fully utilised.

The yearly revenue of most smaller vessels is, however, mostly based on different fisheries where longlining with multifilament gear is used for only part of the year, combined with other methods of fishing such as monofilament longlining, gillnet fishing or jigging. In these situations with only part-time longlining with gear that can be used with mechanised systems, it is very difficult to justify the investment in equipment for mechanised longlining. The random baiter systems were for some time regarded as a low-cost alternative to fully mechanised systems for smaller vessels. However, despite fairly high popularity during the 1980s, the random baiters have never become permanently established in the longline market. The main reasons for this seem to be the poorer quality of baiting, and the fact that most of the gear handling work still has to be done by hand.

Another major longline fishery that is still conducted with manual gear handling and baiting is oceanic longlining for tuna. Despite the fact that this fishery is based on large vessels with high yearly earnings, mechanisation has not been introduced. This is because of the wide hook spacing (50 m). So even if the tuna longliners haul twice the length of line as compared with an autoline vessel (100 vs 50 km), the number of hooks is relatively restricted (3000 vs 20 000–40 000). The work of gear handling and baiting in the tuna fishery therefore do not justify investment in mechanised longline equipment such as baiting machines.

Besides the factors mentioned above, mechanisation may not be introduced for a variety of other reasons. For example, the sizes or shapes of the hooks used in specific longline fisheries may be beyond the range that can be handled mechanically. Likewise, several bait types (e.g. sardines, shrimp or crabs) cannot be used with mechanised systems. Longline fishing based on monofilament gear also falls outside the scope of mechanisation – so far. However, there is continuous development work on more versatile systems for mechanised longlining, and it is to be expected that automation will increase in longline fisheries that are presently based on manual handling and baiting of the gear.

3.5 Longline vessels

The longline is a very versatile gear that can be operated from almost any size and type of vessel. In principle the same longline can be used from a dugout canoe or small rowing boat as the one used on board a 60-m ocean-going autoline vessel, the only difference being the amount of hooks and line carried and operated. The main characteristics of a longline vessel that distinguish it from other fishing vessels are the rail roller, the longline hauler and in particular the setting chute on traditional longline vessels or the baiting and gear handling system on mechanised longliners. Longline vessels may be classified by size (LOA) into smaller (8–15 m), medium-sized (15–25 m) and large (25–50 m) longliners.

Small longline vessels

The smaller longline vessels normally fish in sheltered waters or on fishing grounds near the coast. With few exceptions the operation is based on baiting and gear maintenance ashore, with between one and three crewmembers making 1 to 2-day trips from the land base. Typical examples of smaller longline vessels are shown in Fig. 3.25: typical

(a)

(b)

Fig. 3.25 Small longline vessels: (a) Eastern Canada, (b) Japan, (c) Norway.

(beamy) wooden, coastal longline boats from Nova Scotia; a Japanese wooden coastal longline boat; and a Norwegian steel boat for coastal longlining. Wood remains an important building material for smaller fishing vessels. However, an increasing proportion of newly built vessels are made from fibreglass, steel or aluminium.

(c)

Fig. 3.25 Continued.

Medium-sized vessels

Longline vessels of 15–25 m LOA may fish the coastal or inshore waters along with the smaller longliners, but also have a wider range, fishing the coastal banks and making trips of 2–7 days. The crew may number between three and eight, depending on the mode of operation. With land-based baiting and gear maintenance, a 20 m vessel may carry 90–100 tubs of gear (30 000–40 000 hooks) which are fished during a 2-day trip with a crew of four fishermen. While the vessel is at sea, a team of four to six conducts the repair and baiting of a second set of gear, which is exchanged with the hauled gear when the vessel calls port to unload the catch. Onboard manual baiting is still practised with medium-sized longliners, which then operate with a larger crew. Medium-sized vessels are typically equipped with engines of 250 to 600 HP, giving a maximum speed of 8–10 knots (4–5 m/s). A typical Norwegian medium-sized longliner is shown in Fig. 3.26. This is a 20-m sheltered wooden vessel, with main deck and all superstructures of aluminium.

Large longline vessels

Larger longline vessels are mainly designed for oceanic deep-water longlining based on self-contained trips of more than 3 weeks' duration.

Fig. 3.26 Typical medium-sized Norwegian longline vessel.

Figure 3.27 shows a typical large Asian tuna longliner such as is used in Japan, South Korea and Taiwan. The size of these vessels ranges from 40 to 60 m in length (300–400 tonnes), and they have 20–25 crew-members. Today it is common for trips to be of 18 to 24 months' duration between calls to the home port, and unloading of the catch and refuelling are done in foreign ports or by direct transfer to carriers at sea every 2 months.

The development of modern offshore longlining in the north Atlantic started after the Second World War. Figure 3.28 illustrates this development by showing a sequence of succeeding vessels of the same shipowner (or family of longline fishermen) based near Ålesund on the Norwegian west coast. Basic data on the vessels are given in Table 3.2. The first vessel A was originally built as a wooden sailing vessel around year 1900, but later converted several times. Figure 3.28 shows her appearance after rebuilding in the 1930s, a basic shape that was kept until she was replaced by a new wooden vessel B in 1961. Vessel A was used as a combination vessel, gillnetting for herring and cod in winter and spring and longlining for tusk and ling on the coastal banks and the continental slope during summer and autumn, mainly based upon onboard hand baiting of the lines during 1- to 2-week trips. However, as early as the mid 1930s longer trips were made to Icelandic and Faroese waters, typically landing catches of 30 tonnes of salted fish. The next vessel B was actually a late representative of a large number of wooden vessels built in the period between 1950 and 1965, being a larger and more sophisticated development of its predecessor. However, the changes in basic equipment and mode of operation were not significant,

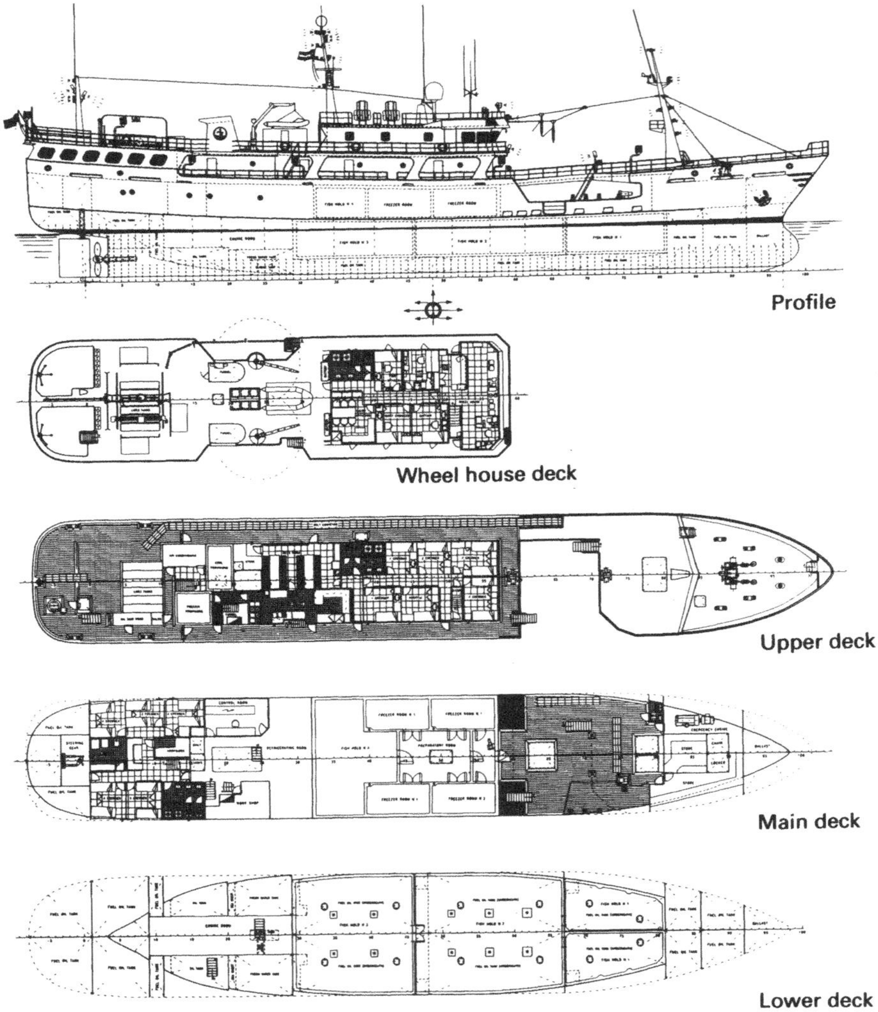

Fig. 3.27 Tuna longliner (after brochure by Astilleros Gondan S.A., Spain).

although this type of vessel allowed a trend towards longer trips to more distant waters. The next vessel C represented a marked shift as this was a steel vessel. The size is further increased and there is specialisation towards a more pure longline operation. The first three vessels were used for longline fishing based on manual baiting of the lines on board. The equipment was basically the same, with the hydraulic hauler as the only mechanised device for handling of the gear. With vessel C, a new development was introduced, namely freezer holds. On this vessel there was only a limited freezing capacity, mainly for keeping the bait, and most of the catch was still kept on ice or salted. With vessel D,

Table 3.2 Longline vessels of the Holmeset family, from 1933 to 1996 (personal communication Mr Per Holmeset, skipper/owner, F/V *Geir*)

Vessel	Year	LOA (m)	Cargo (mt)	Engine (HP)	Other
A	1933	19	30	100	Wood, hand baiting
B	1961	22	50	200	Wood, hand baiting
C	1968	27	80	400	Steel, hand baiting
D	1978	32	155	700	Steel, autoline, freezer
E	1989	38	300	1100	Steel, autoline, freezer, factory

significant new developments were made. Even though the Mustad Autoline system had been introduced some years earlier, this was one of the first newly built vessels that had an automated longline system installed. A construction that is common nowadays, the shelterdeck, is found on this vessel, and improves the working conditions and the security of the fishermen significantly. The freezing capacity is increased to keep larger parts of the catch frozen. The latest vessel built for this owner so far E, still represents the most modern longline vessel design fishing in the north east Atlantic. With a length of 38 m and a beam of 8.5 m she measures 172 gross tonnes. The 1100 HP diesel engine gives a maximum speed of 12.5 knots (6.3 m/s). Other main characteristics of the vessel are as follows:

- Below main deck: Engine room, freezer hold (176 m^3), refrigerated hold (170 m^3), freezer storage for bait (37 m^3)
- On the main deck: Storerooms, hydraulic machinery, freezer tunnel, vertical plate freezers, gear storage, hauler room, autoline system with 40 000 hooks, fish bins with hydraulic regulation of the bottom for easier handling of the fish, fish processing machines for de-heading, splitting and fillet production
- On the shelterdeck: Crew accommodation for 16, mainly in single cabins; galley, mess room, bathrooms
- Pilot house: Advanced electronic equipment for navigation, fish finding and communication

As shown in Fig. 3.28, the pilot house is located in the aft part of the vessel. This is historically a traditional design for Nordic fishing vessels. However, in recent years, an increasing number of fishing vessels are being built with the pilot house at the front. On larger longline vessels the pilot house is still aft, because this gives the skipper the best opportunity for monitoring the line during hauling. For improved manoeuvrability most larger longliners are now being fitted with a Becker rudder. Figure 3.29 shows technical section drawings of a longline vessel similar to vessel E.

The vessel designs shown in Figs. 3.28 and 3.29 are good representatives of the modern mechanised longline vessels that have been

Fig. 3.28 Development of modern Norwegian ocean-going longline vessels.

developed over the last 10 years as self-contained units for longline fishing in the north Atlantic and north Pacific groundfish fishery. The latest development is a series of large longline vessels built for fishing Patagonian toothfish at large depths (1000–2500 m) off Argentina. The basic vessel design is as shown in Fig. 3.29, but these are considerably larger vessels (50–55 m). An interesting new development with the most recent of these vessels is a so-called moon-pool, a 3 × 4 m vertical hole through the centre of the vessel through which the lines are hauled

(Fig. 3.30). The main advantages expected from this arrangement are less loss of fish during hauling, and improved working conditions and increased safety for the deck crew at the hauling position.

3.6 Electronic instrumentation

Before the Second World War, the instruments on fishing vessels were very basic and simple, often limited to a compass, a towing log, a clock and a barometer. In the post-war period there has, however, been a considerable development of a variety of electronic equipment for navigation, fish finding and communication. Information is a prerequisite for successful fishing, both stored information (experience) and new, current information. Modern electronic equipment and information technology have therefore contributed significantly to the increased efficiency of fish capture during the last 50 years. This is also the case with longlining. Electronic equipment used in longline fishing is described below.

Fish-finding equipment

Today, effective purse seine and trawl fishing strongly depend on fish finding by the use hydroacoustic equipment such as sonars and echo sounders. Fish capture with a passive gear like longline is based more on the fish finding the gear than on the gear 'finding' or chasing the fish. The importance of fish-finding equipment is relatively low when using passive (and in particular baited) gears. With demersal longlining, echo sounders are mainly used to record the desired depth and partly to evaluate adequate bottom substrate and configurations for setting the gear. This is mainly because it is difficult to distinguish fish close to the sea bed from the bottom echo, as the fish is 'hidden' in the so-called acoustic dead zone. However, with relatively high fish densities and with fish distributions slightly above the sea bed, echo sounders provide useful information also for demersal longlining. With semipelagic longlining, the echo sounder is a more important tool, not only for fish finding but also to position the lines at the correct depth during setting. The bottom depth may for instance vary considerably, while the fish concentrations are found at a certain depth layer. By judging the variation in fish to bottom depth, the skipper can advise the crew on the length of sinker/float lines to be used, so as to match the gear depth to the fish depth as closely as possible. A similar practice applies to pelagic longlining, where echo sounding provides information that enables the lines to be suspended at depths matching the fish registrations.

In pelagic longlining for species such as tuna, temperature is an

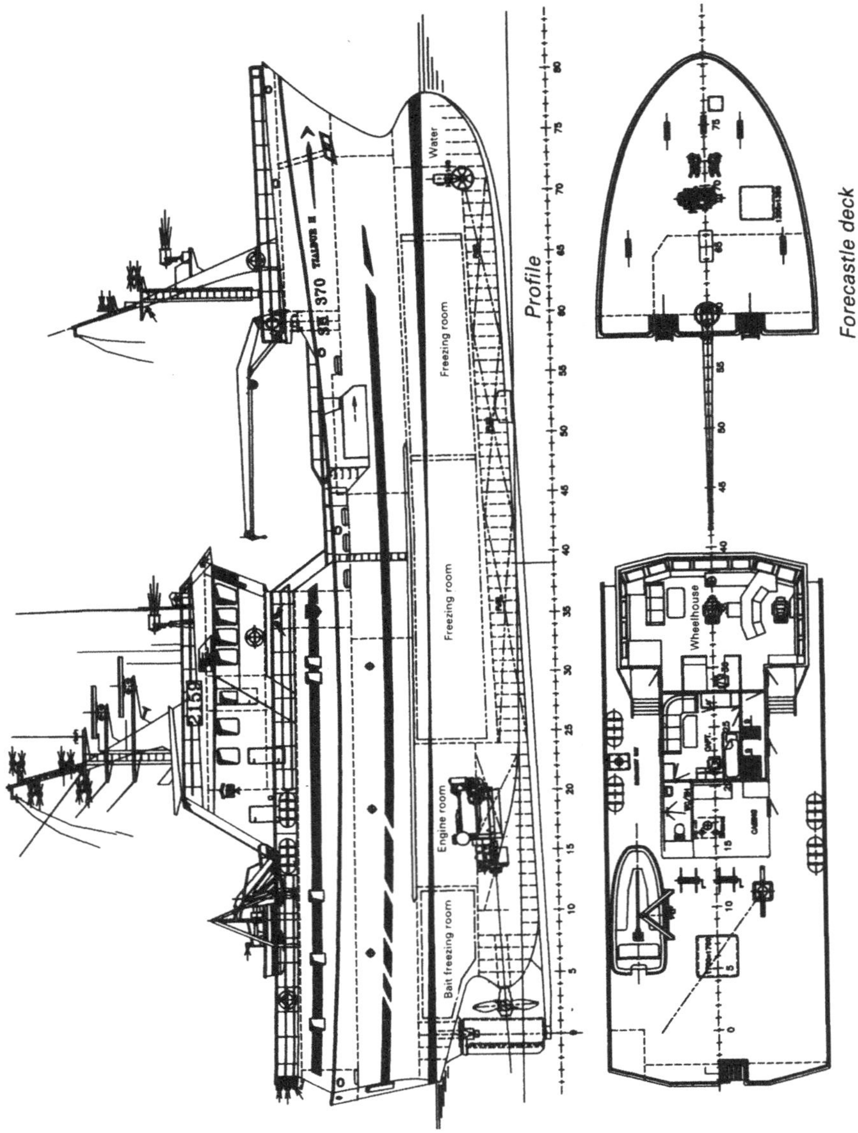

Fig. 3.29 Modern longline vessel: section drawing (after drawing of Selstrand Shipyard, Norway).

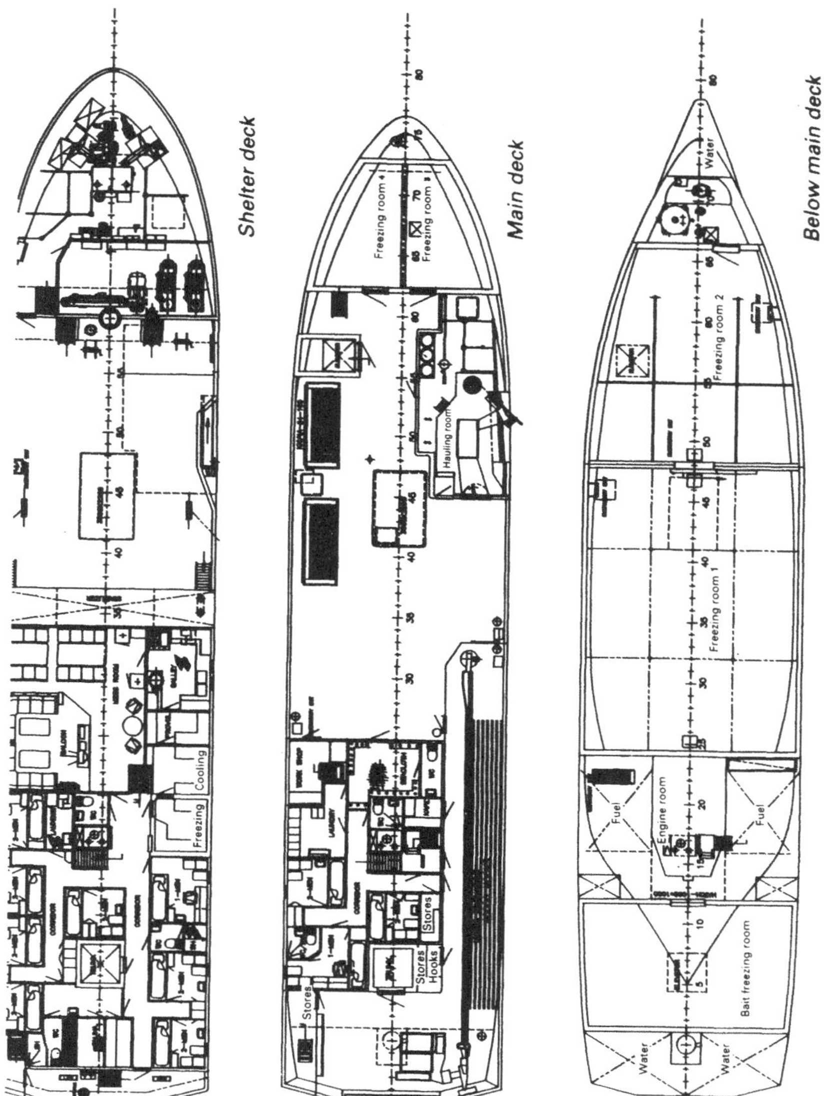
Shelter deck
Main deck
Below main deck
Cooling
Freezing
Freezing room 4
Freezing room 3
Hauling room
Stores
Hooks
Stores
Stores
Water
Freezing room 2
Freezing room 1
Engine room
Fuel
Fuel
Bait freezing room
Water
Water

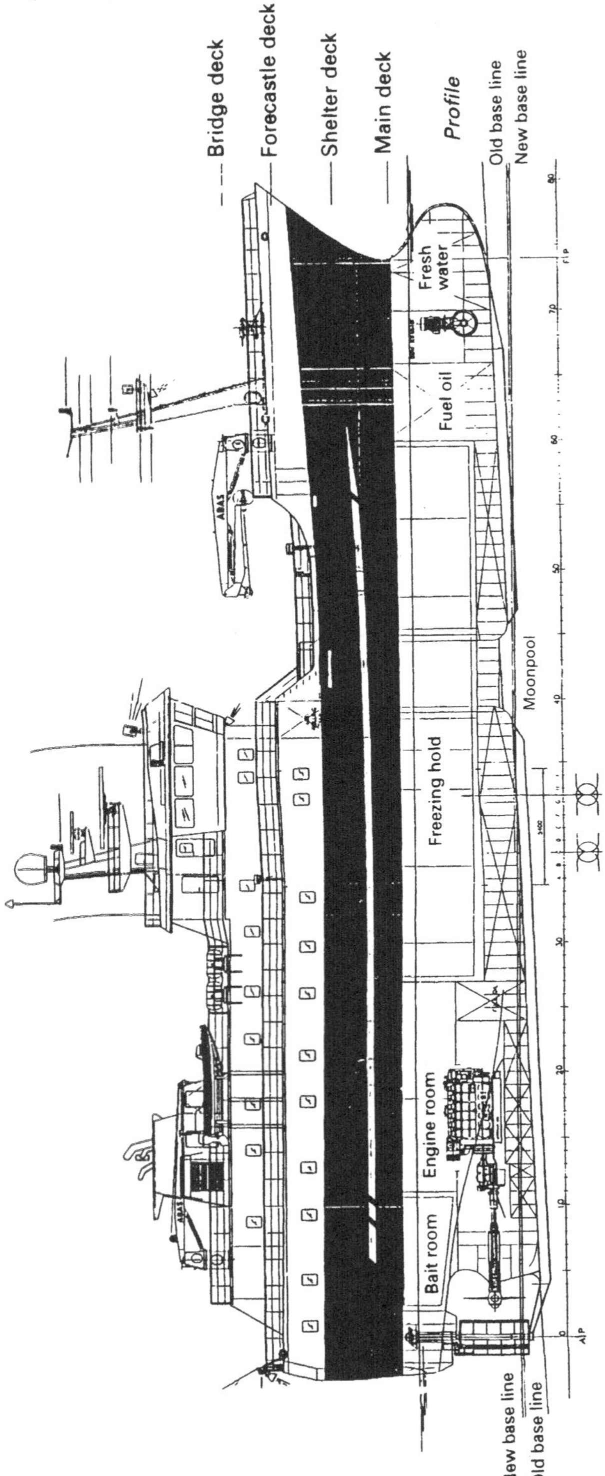

Fig. 3.30 Longline vessel with moonpool (after drawing of Brattvåg Shipyard, Norway).

important parameter as the distribution of the fish is closely related to certain temperature preferences. Tuna longliners are therefore commonly equipped with sensors and display units for continuous temperature recording. In addition, different tuna fleets are equipped to receive regularly updated satellite-based temperature charts, showing the surface temperature distribution for large oceanic areas. These services, which are provided by the fleet management or a governmental agency in the home country, have clearly improved fishery efficiency.

Navigation

Besides the general navigational needs, precise navigation is an important success factor in longline fishing, for both fish finding and operational efficiency. Particularly with demersal longlining, fishing success depends strongly on setting the gear accurately at positions where good catches have been obtained. In coastal fishing this has traditionally been solved by the use of visual bearings of characteristic landmarks, a method that is still used. However, in offshore fishing and also in small-scale inshore fishing, there has recently been an increased use of accurate electronic positioning equipment. Land-based navigation systems such as those of Decca and Loran have been superseded by the more accurate, easy to use and affordable satellite-based positioning systems. This development has significantly improved the longline fisherman's abilities to map and recover favourable fishing spots. The development of electronic position plotters and electronic charts has improved this further. As these are linked to the satellite navigation system, the position data of individual longline fleets can be easily stored. This provides easy recovery of marker buoy positions, which clearly improves the catching efficiency as any time used searching for buoys is wasted, non-productive time that could be used for hauling gear. Thus, when a fleet of gear is retrieved, the modern longline skipper can just select the position (way point) of the next end buoy he wants to recover from the electronic chart or position plotter. The display then shows the distance and bearing to the buoy position. As the automatic pilot is linked to the navigation system, the skipper just has to set the desired speed and the electronics will automatically take the vessel to the marker buoy. Along the course, the display shows the remaining distance between vessel and buoy. If desired the skipper can also set an alarm function that sounds when there is a set distance or time left (e.g. 0.2 nautical miles or 5 min) before reaching the buoy. If the visibility is good, the buoy should at this point be easily spotted. However, in the dark or in foggy or dismal weather another piece of navigational equipment, radar and strong searchlights, are used to find the buoy. Electronic navigation equipment has thus significantly improved the efficiency of longline fishing, both by making a higher catch possible through the exact positioning of the gear, and through increasing productivity because little time is wasted searching for the marker buoys.

The radio bearing finder is another navigation device that has contributed to improved longline efficiency. Particularly when the fishing is poor, it is important for the skipper to obtain information on the fishing success of other vessels when making a decision to change the fishing ground. He is usually well informed about other fishing vessels that are within a reasonable range, among which there is normally daily radio communication. Radio silence by a vessel often indicates that this vessel is making good catches and is not using the radio transmitter to avoid being located by the radio direction finder of neighbouring vessels. Some skippers therefore carefully record radio bearings of other vessels when they are transmitting in order to obtain a good overview on their relative positions and movements.

Communication

The normal situation is, however, one of frequent communication between vessels and between vessels and shore. Modern smaller longline vessels are normally equipped with a walkie-talkie, a VHF-radio telephone and/or a cellular telephone. In addition to the communication systems mentioned, larger high-seas longline vessels may use some or all of the following communication systems: medium-wave and short-wave radio transceivers, satellite telephones, telex and fax machines. As mentioned above, good communication systems are important for obtaining information on the fishery in the area as a basis for improved decision making when choosing fishing grounds. In addition, good communication between vessel and shore may increase profitability, for instance by providing the latest market prices so that the highest priced fish species can be targeted in particular or by providing technical advice so that broken equipment can be repaired at sea thereby avoiding fishing time being wasted to call port for repairs.

Internal communication systems have also improved the longline operation. Particularly on larger vessels, where the skipper does not have direct visual or oral communication to different working positions on board, internal loudspeaker and/or video systems have proved to be very useful for smooth operation. Normally such audio-visual communication and monitoring systems are located at the hauling, maintenance and setting positions to keep the skipper updated on possible operational irregularities. Likewise, different monitoring and alarm systems for the engine are normally found on the bridge so that immediate action can be taken to avoid breakdowns of the sophisticated engine systems of modern longline vessels – breakdowns that may be very costly and time consuming and may even be fatal.

Integrated productivity information systems

A large number of different electronic instruments are found on the bridge of a modern fishing vessel. They are normally separate units with little interlinkage and without comprehensive collection and storage of data. In longlining such comprehensive data collection would be particularly useful, as capture success is based largely on the evaluation of historic information. Further, longline fishing is somewhat similar to an industrial assembly-line production, with the hooks and fish being hauled one by one. Systems to store data on relative productivity would therefore be beneficial. A recently developed Icelandic system (Line Tec 2000) meets some of these needs. By linking existing instruments and new sensors it may prove to be a very useful tool for increasing productivity in longline fishing. The new approaches employ sensors for measuring the length of line hauled, hauling speed and tension of the line during hauling, as well as a catch (fish counting) sensor. The tension meter tells the skipper when he should take action to avoid breakage of the line so that the gear is not lost or time wasted on the recovery of intermediate buoys or the other end marker buoy. The length of line and the number of fish per fleet are automatically stored in a PC on the bridge, together with position, depth and time data. Data on fish species, size and other comments can be added separately. Thus, the skipper can gradually create his own database, with catch rates obtained on different grounds and times of year: information that has hitherto been carefully gathered in log books and in the mind of the clever longline skipper. Creating a computer data-base provides new and valuable possibilities for scanning historical catch rate and position data. There is every reason to believe that this and even more sophisticated databases will be important tools for successful longline fishing in the future.

3.7 *Safety aspects of longlining*

In general, seafaring and fishing imply a relatively high risk of injury and loss of life as compared with most other professions – even though the risk level has been significantly reduced through improvements in vessel technology, safety and rescue equipment and communication systems. Today most larger longline vessels have a shelter deck, which provides a weather-protected work environment for most of the deck crew. The only working position that is still exposed is at the side hatch when hauling or setting the gear. In bad weather, large water masses can flow in through the hatch, and on some occasions longliners have capsized and sunk because the drainage pumps have not been able to cope with the water flowing on to the main deck. To improve this weak point, side hatches have been equipped with hydraulic closing mechanisms that can be activated both from the maindeck and from the bridge. Also, the hauling position of most larger vessels is now normally built in. Possible water inflow is thus limited to the relatively small volume of this separate hauling room, so that large water volumes on the main deck are avoided. If the hauling of longlines through a 'moonpool' proves to be a successful solution (see Fig. 3.30), this construction may further reduce the risk of water inflow.

Setting the gear represents another specific high-risk part of the longline operation. Although serious accidents are rare, there is always a risk of fishermen being pulled overboard if caught by the outgoing mainline or buoy line, or of being injured if hit by outgoing hooks during setting.

3.8 Handling of the catch

Longline fishing provides ideal conditions for excellent handling of the catch, as fish are hauled in one by one. In general, the fish is gaffed by the fisherman at the rail roller and normally bled at once by a cut in the throat before it is dropped in the bleeding bin for about 30 min to drain most of the blood before further processing. From this point there are different ways of handling the catch. With shore-based longline operations, the fish may be left undressed in the hold if the time between capture and unloading is short (a few hours). With somewhat longer trips (1–3 days) the fish is dressed and stored on ice in bins or boxes in the hold. A recently developed technique for use on fishing trips lasting 1–3 days is to store the fish in containers of about 1 m^3 with crushed ice and sea water. This has been found to give better quality than traditional icing, and saves labour at sea and during unloading as a few containers contain the same catch as earlier was stored in several hundred boxes.

With longer trips the fish has to be dressed before it is stored with ice in the hold. It used to be common practice to store the fish on ice up to 4 weeks from the time of capture. The traditional alternative to preservation on ice has been salting. Then the fish is normally dressed and split (backbone removed) before being stacked in bins in the hold.

As mentioned earlier, frozen storage of the catch is getting more and more common, particularly on large ocean-going vessels. After dressing and washing, the fish is frozen, usually in vertical or horizontal plate freezers into blocks of about 40 kg that are stored in the freezer hold.

In recent years there has also been a clear trend towards further onboard processing of the catch, with the use of machines for de-heading, splitting and filleting of the fish. This gives the skipper the possibility of increasing the catch value by processing different fish species to products that give the best price.

The fish handling procedures described above are mainly those used for groundfish handling. In the longline fishery for tuna, careful handling of the catch is of particular importance for a high quality and hence highly priced product. The fish is killed as quickly as possible using a tool to damage the spinal cord. The fish is then shock frozen and stored at −40°C. Tuna that is processed in this manner is called *shime-shime* tuna, meaning very high quality tuna that obtains a very good price in the Japanese *sashimi* market (*sashimi* being a traditional dish in Japan).

In some longline fisheries the catch is stored alive in wells with circulating sea water, as in the Japanese coastal longlining where certain fish species fetch particularly high prices, if marketed live.

3.9 The longline fisherman

The fishing profession has traditionally been conducted by men, as denoted by the occupational term 'fisherman'. This is particularly the case in the longline fisheries, which are regarded as one of the most hard and laborious ways of fishing. Although a few women may be found as crewmembers and even skippers on longline vessels, these must still be regarded as the few exceptions that confirm the general rule of longlining being a heavily male-dominated occupation. However, this is with respect to the actual fishing operation and vessel-based part of the longline fishing. In the traditional hand baiting fisheries, the onshore baiting of the lines may be dominated by women, as this work often represents one of the few land-based employment opportunities in rural fishing communities.

The preceding text has mainly concentrated on the principles and technical descriptions of longline fishers, and little has been said about the conditions and daily life and routines of the longline fisherman. Just as longline fisheries vary greatly, so do the conditions for the longline fisherman – from the coastal fisherman conducting daily fishing trips when weather permits to the crewmember of an Asian tuna longliner who may spend as long as two years on the vessel before returning to the home port. Although it is far beyond the scope of this book to give a comprehensive description of the occupational and social conditions of longline fishermen, a few examples are needed to illustrate the daily life of the people involved in different longline fisheries.

Small-scale coastal longlining

The small-scale longline fisherman often operates single-handed at sea and may also be alone doing the baiting and gear maintenance work. He is usually an early riser, either to start baiting the lines to be ready for sea as soon as a sufficient amount of gear has been prepared and weather permits him to leave port, or to leave early for sea with pre-baited gear. The small-scale longline fisherman is thus often found in the early morning steaming to the fishing ground with his 25–30 ft (7.5–9 m) vessel, to shoot the 2000–4000 hooks of prebaited gear. When the fishing position is chosen and the gear is set, he usually stays at sea waiting the 1–4 hours necessary for the fish to get hooked before he starts to retrieve the gear. Then he has the opportunity to have a snack and a cup of coffee, usually prepared at home before leaving. During hauling, he manoeuvres the boat, de-hooks and bleeds the fish, replaces tubs as they are filled with retrieved gear with empty ones under the automatic hauler and does everything necessary to ensure a smooth hauling operation. When the hauling is finished, he heads for port. He will often use the steaming time to dress the fish. When he is back in port the catch has to be delivered, the tubs with retrieved gear have to be taken ashore for rebaiting, and the boat has to be cleaned before he can call it a day and look forward to another day of repairing and rebaiting the gear.

The single-handed longline fisherman has some clear advantages. He is independent and can to a large degree design his own day and pace. He can also have a fairly normal family and social life, being at home most nights, weekends and holidays. On the other hand, operating from a small vessel he might experience long periods without income as bad weather may prevent him from going to sea. If he is doing the baiting himself, he might also lose valuable good fishing days ashore to rebait the gear. The small-scale fisherman may also experience great yearly variations in income caused by changes in fish availability. Due to the limited range of his small vessel, he is highly dependent on near-shore fish densities. When this is the case he might experience very good fishing whereas with a more offshore fish distribution the fishing can be really poor. His worst and toughest days are those when weather conditions are medium to bad but not bad enough to force him to stay in port. Then the working conditions may be extremely difficult and exhausting, with the small boat rolling and pitching heavily. Finally, the safety aspects are not the best with the one-man longline operation. The improved quality of small-scale fishing vessels and the availability of sophisticated navigation and safety equipment and good communication systems have improved the safety of single-handed fishing. However, there is always an extra risk when operating alone at sea.

Larger coastal vessels

Compared with the one-man operation, intensity and specialisation are increased for fishermen on a larger coastal longliner (15–25 m LOA).

The lines are still baited ashore, but in this fishery we commonly find a separation between fishermen and baiters or shore crew. In some arrangements, it may be common to rotate jobs, so that the crewmembers alternate between sea and shore duties. The intensity of this fishery is only limited by fish availability and bad weather, as there is normally a steady and adequate supply of pre-baited gear. A typical coastal longline vessel leaves port with four crewmembers and 20 000–30 000 hooks of pre-baited tub gear. During the 2- to 8-hour steaming to the fishing ground, there is a good opportunity for the crew to get some rest except for their watch at the helm, a duty that normally rotates between the crew. Professionally the crew often consists of a skipper, a combined mate and engineer and two fishermen. During setting, the skipper is always found in the pilot house. The others are working on deck, normally two of them switching full for empty tubs at the setting chute and attaching weights or weights/floats to the gear. The third fisherman on deck takes care of the anchors, buoys and buoy lines. When the gear is set, there is a short break with the opportunity to get a simple meal and a short rest. The food preparation is normally very simple and is either allocated to one of the fishermen (often the junior) or done on a self-serve basis.

During hauling, the skipper is usually found in the pilot house manoeuvring the boat. However, when the fishing is good and the weather is reasonable, he might join the others on deck and assist either at the rail roller, taking care of the fish (dressing, boxing, icing) or storing tubs of retrieved gear, while the manoeuvring is done by the man at the rail roller by extended steering and engine controls. The hauling usually takes 24 hours or more, so the crewmembers normally have a 2-hour break divided sequentially among the four during the hauling period. When the last anchor is brought on board, hatches and loose items are secured before the course is set for port and there is a new opportunity for three of the men to relax while the fourth crewmember watches the helm. When in port the crew normally participates in unloading the catch, changing the retrieved gear for baited lines and usually getting the opportunity to have a hot meal ashore before they head back to sea.

In this way the working week of a longline fisherman on a larger coastal vessel may include two or three fishing trips with close to 100 working hours. They normally have the Sunday off, with the opportunity to recuperate before another tough week at sea. The fishing on medium and large longline vessels is more intensive compared with the small boats, and the larger range and better living conditions on the larger boats also often mean that the fishermen spend longer periods away from home. However, this is usually compensated by fairly long breaks (of several weeks) in between different fishing seasons, particularly if the fishing has been good so that the crew feel that they can afford this and still have a sufficiently high yearly income.

Large-scale offshore longlining

On the largest longline vessels with trip durations of 3–7 weeks, the specialisation amongst the crew is further developed. With respect to operation intensity, fishermen on the larger offshore operating vessels also work long days and weeks at sea, but often experience more regularity and a more orderly working rhythm as compared with their colleagues on smaller coastal vessels. A typical Norwegian autoline vessel with no further fish processing than block freezing of the dressed fish, normally has 12 crew members: a skipper, an engineer, a cook, two fishermen for handling and processing the catch and seven fishermen for handling the gear. Among the fishermen there will be at least one who has the proper navigation licences to formally act as stand-in for the skipper if needed. The skipper normally works only in the pilot house, where his main task is to supervise the whole fishing operation with the best possible outcome as his objective. He decides where to set the gear and spends most of his working hours manoeuvring the vessel during setting, hauling or steaming between fleets of gear or different fishing grounds. He will typically work the watches from 06.00 to 12.00 and 16.00 to 24.00, which gives him a 4-hour break in the afternoon, 6 hours' sleep at night and a total of 14 working hours daily. The engineer is normally the person who relieves the skipper in the pilot house, first from midnight till 06.00, then from noon till 16.00. Leaving the pilot house in the afternoon does not mean that he is off duty. He then goes straight on to a 4-hour watch (16.00–20.00) as a real engineer. During that period he has to keep up with his duties in the engine room, doing the periodic checks on the main and auxiliary engines and the electric and hydraulic systems, and maintenance work on technical installations throughout the vessels, including the baiting machine. The cook is the only person on board working a day shift as he is on watch in the galley from 06.00 until 22.00. The two fishermen working with the handling of the catch usually work together and they normally follow the same watch scheme as the skipper (06.00–12.00 and 16.00–24.00). They do all the handling of the fish after it is bled, which includes dressing, washing, freezing or icing. As there is no one doing this work in their absence, they have to start their watch with a large number of undressed fish. If the fishing is extremely good, fish processors may have to break their watches (that is, work in their off-duty time) and may even need to be assisted by some of the other fishermen. However, this extra work is normally done without complaint as high catch rates mean better income. The remaining seven fishermen go on a watch scheme with 8 hours on, 4 hours off, 8 hours on and 8 hours off watch. This adds up to a weekly workload of 96 hours on duty.

Of the seven fishermen handling the gear, four are on watch at the same time. When hauling (18–20 hours per day), one man is at the rail roller with gaffing and bleeding the fish as his main occupation; one man is reshaping hooks; and the last two men are mending the gear (replacing broken snoods and hooks) and getting the lines properly stored on the rack magazines ready for setting. During setting there are normally three fishermen on deck: one to feed bait to the baiting machine, one to arrange the racks in proper order and to check that the hooks slide easily off the racks into the baiting machine, and one fisherman to arrange and throw anchors, buoy lines and buoys. This means that the surplus man of the four on deck watch can have an extra break during setting of the gear. Within the deck watches it is usually so arranged that the first four hours of a watch are spent on gear maintenance and the last four are spent either at the rail roller or re-shaping hooks, often rotating between these two duties so that the time at the rail roller is not more than one hour.

Most fishermen work hard and long hours under relatively tough working conditions. Despite mechanisation and other improvements, longlining is still regarded as one of the most laborious fishing methods. However, in the case of the north Atlantic autoline fishery there has recently been a trend towards a regular crew shift arrangement based on one and a half crew so that the fishermen spend two trips at sea and then have one trip off. The general fisherman still puts in about 3000 working hours per year during an average of 200 days at sea, which is twice the effort compared with industrial workers on shore. Because of the hard working conditions there has also been a trend towards younger people as crewmembers on longline vessels, the majority of whom have families and children. The shift system, providing regular breaks from sea duty, makes it easier to combine the fishing profession with a family life. However, considering the high working effort in longline fishing and even better shift arrangements in comparable professions (such as the merchant navy, the offshore oil industry etc.), future crew recruitment to the longline fleets probably depends on increased profitability that would enable vessels to afford two sets of crew with a 1 : 1 off–on duty arrangement.

Chapter 4
Main Target Species and the Relative Importance of Longlining

Fishing takes place in all oceans, but most of the fishing activities are concentrated along the coasts and on the continental shelves and slopes where fish are most abundant. Today nearly all marine stocks are fully exploited and several are overexploited. The total annual landings from marine waters were 81.7 million tonnes in 1991 of which fish constituted 69.2 million tonnes. In terms of quantity, the pelagic stocks (clupeids (e.g. anchovy, sardine, herring and menhaden), Alaska pollock, mackerel and capelin) are the most important marine fish resources. These stocks are caught mainly with trawl and purse seine, which are the most efficient and economic ways of catching schooling fish that occur in high densities. Fish swimming in dense schools are not efficiently caught by longlines because the sizes of longline catches are limited by the number of hooks. In addition, the cost of bait represents a significant share of the total expenses in longlining. The earnings per baited hook of catching these small, low-priced fish are therefore too low to make longlining profitable.

However, fishing vessels using longlines are common on all oceans targeting highly priced tuna and demersal fish species which are more valuable. It is beyond the scope of this book to cover all these fisheries. This chapter will describe the main fisheries where a significant proportion of the catches are taken with longlines:

- tuna in the Atlantic, Indian and Pacific Oceans
- groundfish in the north Atlantic
- groundfish in the north eastern Pacific
- hake in the Atlantic.

4.1 Tuna in the Atlantic, Indian and Pacific Oceans

A large fleet of tuna vessels is operating with pelagic longlines in the Indian Ocean and in the central and southern Atlantic and Pacific Oceans. The fishing method was perfected by the Japanese in the 1930s, but its use did not spread outside Japan until after the Second World War. Currently as many as 2000 longline vessels are engaged in this fishery (primarily for tuna), and the home ports of most of these vessels are located in Japan, Taiwan and South Korea. Large, specialised vessels are used in this fishery, and they stay away from their home ports for 10–24 months. Pelagic longlines with long snoods and wide hook spacing are used (see Section 3.1). The catch is handled very carefully on board and frozen at −40°C to −60°C to fulfil the demand for high quality fish. Their catches are landed directly in Japan, where tuna for the *sashimi* market is highly priced. This market has very stringent quality requirements, and to ensure the highest price, longliners must land their catches fresh or stored in super-cold freezer holds. Canned tuna is the other important market for the annual world catch of about 400 000 t of longline-caught tuna.

Longline and purse seine are the most important gears in the Japanese and Korean tuna fisheries (Tables 4.1 and 4.2). The total annual catches of tuna landed by Japanese and South Korean vessels were 312 703 and

Table 4.1 Mean annual catch (tonnes) of tuna by Japanese vessels, 1988–92, by major gear types ('other' is mainly pole and line, and drift net). The percentage of the catches taken by different gear types is given in brackets.

Species	Longline	Purse seine	Other	Total
Bluefin	11 484 (68)	3 691 (22)	1 800 (11)	16 975
Yellowfin	50 108 (45)	47 713 (43)	12 555 (11)	110 376
Albacore	17 059 (38)	2 496 (6)	24 920 (56)	44 475
Bigeye	128 517 (91)	3 729 (3)	8 631 (6)	140 877

Source: Japanese Ministry for Agriculture and Fishery.

Table 4.2 Mean annual catch (tonnes) of tuna by South Korean vessels, 1988–92, by major gear types ('other' is gillnet, and squid angling and gillnet). The percentage of the catches taken by different gear types is given in brackets.

Species	Longline	Purse seine	Other	Total
Yellowfin	19 521 (32)	38 339 (63)	3 075 (5)	60 935
Albacore	3 414 (90)	– (–)	377 (10)	3 791
Bigeye	29 616 (100)	74 (0)	10 (0)	29 700

Source: Agriculture and Fisheries Statistics Bureau, South Korea.

98 154 t (means for 1988–92), respectively, of which yellowfin and bigeye tuna constituted the largest proportions. The fishery for bluefin tuna is monopolised by the Japanese fleet. Yellowfin tuna is caught both with longline and purse seine, whereas bigeye tuna is almost exclusively caught by longline. Bigeye tuna is distributed deeper in the water column (about 300 m) than yellowfin tuna and albacore (which both occur at a depth shallower than 200 m) and cannot be reached with purse seine. Standard longlines are set at a maximum of about 170 m. In the 1970s a 'deep longline' was developed to target bigeye tuna. In Taiwan, longline is the traditional gear used to catch tuna, but development of a purse seine fishery began in the early 1980s. The mean annual Taiwanese longline catches (1988–92) of yellowfin, albacore and bigeye tuna were 18 564, 44 926 and 20 843 t, respectively.

The Japanese longline fleet mainly targeted yellowfin tuna and albacore until about 1970 to supply the growing US canning industry. Then the fleet gradually switched to bigeye tuna, which currently accounts for more than 60% of the catch (Table 4.1). This shift was driven by economic and market forces. The substantial increase in fuel and labour costs in the 1970s forced the Japanese vessels to reduce their export to the low-value canning market, and to emphasise production of tuna to supply the *sashimi* market in Japan where bigeye tuna fetches a higher price than yellowfin tuna and albacore. When the Japanese fleet switched to catching bigeye tuna during the 1970s, the Taiwanese fleet specialised in catching albacore for export to the US canning market where albacore is more valuable than bigeye and yellowfin tuna. More than 50% of the Taiwanese tuna catch is currently of albacore.

Both Taiwan and South Korea entered the longline industry in the early and mid 1960s, and their fleets were based on used vessels purchased from Japan that modernised their longline fleet during this period. The South Korean fleet initially specialised in the production of tuna for the US canning industry. Later the South Korean vessels concentrated on catching tuna for the more valuable *sashimi* market in Japan. Today the longline fleet in South Korea is nearly identical to the Japanese fleet and predominantly catches bigeye tuna (about 55%, Table 4.2). These changes over time, driven by economic and market forces, show how the longline fleet has been capable of targeting certain species for particular markets.

Catch statistics from the Taiwanese tuna longline fishery show that the mean yearly catch rates (number of fish per 100 hooks) in the Atlantic, Indian and Pacific Oceans have fluctuated between 1.0 and 5.5, with the highest catch rates obtained in the Pacific Ocean and the lowest in the Indian Ocean. From the late 1960s to the early 1990s there has been a clear decrease in the mean yearly catch rates for all three oceans, indicating that the stocks are decreasing. This decrease may partly be due to the increased fishing effort over the last decades.

Taiwan alone has increased the total annual number of hooks fished from 38 million in 1967 to 345 million in 1992.

4.2 Groundfish in the north Atlantic

Fishing demersal fishes with longlines has a long tradition in the northern Atlantic. The longline vessels used in these fisheries vary greatly in the size and number of crewmembers, ranging from small vessels (about 8 m) operated by one man to large autoliners (up to 50 m) with 12–15 crewmembers. The smaller vessels stay at sea for one day, the lines are baited on shore, and the catches are landed fresh, whereas the large vessels, most of which are equipped with mechanised baiting machines, stay at sea for several weeks and land their catches iced, frozen and in some cases processed.

Tables 4.3, 4.4 and 4.5 show that large proportions of several of the most important groundfish resources in the northern Atlantic are taken by longlines. In addition, trawl and gillnet are important fishing methods in harvesting demersal fish stocks in this region. Longlining is a more important fishing gear in Norway than in Iceland and Canada,

Table 4.3 Mean annual landings (tonnes) of groundfish, 1989–93, in Norway by major gear types. The percentage of the landings taken by different gear types is given in brackets.

Species	Longline	Jigging	Gillnet	Danish seine	Trawl	Other	Total
Cod	31 324 (16)	25 262 (13)	62 672 (32)	15 821 (8)	57 875 (30)	1 067 (1)	194 021
Haddock	16 924 (50)	681 (2)	3 019 (9)	3 987 (12)	9 392 (28)	98 (0)	34 101
Tusk	25 207 (90)	367 (1)	1 634 (6)	26 (0)	427 (2)	465 (2)	28 126
Ling	16 160 (76)	70 (0)	4 128 (20)	16 (0)	751 (4)	36 (0)	21 161
Halibut	404 (70)	14 (3)	100 (17)	10 (2)	47 (8)	2 (0)	577
Greenland halibut	2 373 (13)	40 (0)	1 832 (10)	22 (0)	14 548 (77)	148 (1)	18 963
Wolffish	2 021 (74)	78 (3)	140 (5)	95 (4)	379 (14)	7 (0)	2 720
Total	94 413 (32)	26 512 (9)	73 525 (25)	19 977 (7)	83 419 (28)	1 823 (1)	299 669

Source: Directorate of Fisheries, Bergen, Norway.

Table 4.4 Mean annual landings (tonnes) of groundfish, 1989–93, in Iceland by major gear types. The percentage of the landings taken by different gear types is given in brackets.

Species	Longline	Jigging	Gillnet	Trawl	Other	Total
Cod	46 643 (15)	18 462 (6)	64 954 (22)	159 645 (53)	12 810 (4)	302 513
Haddock	8 192 (15)	250 (1)	5 150 (9)	37 263 (68)	3 943 (7)	54 797
Tusk	4 735 (93)	64 (1)	149 (3)	127 (2)	44 (1)	5 118
Ling	1 306 (27)	19 (0)	1 259 (26)	1 667 (35)	549 (12)	4 799
Halibut	641 (44)	23 (2)	32 (2)	619 (43)	137 (9)	1 452
Wolffish	8 564 (57)	133 (1)	132 (1)	4 893 (32)	1 303 (9)	15 026
Total	70 081 (18)	18 951 (5)	71 676 (19)	204 214 (53)	18 786 (5)	383 708

Source: Marine Research Institute, Reykjavik, Iceland.

Table 4.5 Mean annual landings (tonnes) of groundfish, 1987–91, in Canada (east coast) by major gear types. The percentage of the landings taken by different gear types is given in brackets.

Species	Longline	Handline	Gillnet	Trap	Danish seine	Trawl	Other	Total
Cod	50 747 (12)	21 185 (5)	56 719 (14)	48 666 (12)	15 197 (4)	215 368 (52)	2 447 (1)	410 329
Haddock	6 931 (27)	329 (1)	387 (2)	3 (0)	670 (3)	17 334 (67)	41 (0)	25 695
Tusk	3 386 (93)	63 (2)	66 (2)	18 (0)	–	69 (2)	41 (1)	3 643
Halibut	1 831 (79)	40 (2)	81 (3)	–	7 (0)	332 (14)	38 (2)	2 329
White hake	7 636 (53)	214 (1)	3 121 (22)	16 (0)	1 332 (9)	2 106 (15)	54 (0)	14 479
Total	70 531 (15)	21 831 (5)	60 374 (13)	48 703 (11)	17 206 (4)	235 209 (51)	2 621 (1)	456 475

Source: Halliday & Clark (1995).

where trawl is the dominant fishing gear for catching demersal fish. The introduction of mechanised longline systems in the mid 1970s was of particular importance for maintaining a large longline fleet in Norway.

The catches of species such as tusk, ling, halibut and wolffish are mostly landed by the longline fleet (Tables 4.3, 4.4 and 4.5). These species do not occur in high densities, and they stay on the seabed, often on rough bottom where towed fishing gears cannot be operated. Because of their distribution, habitat selection and behaviour, these stocks are therefore caught most effectively and economically by longlines.

Tusk and ling are found on the same fishing grounds, and are caught by longlines in a mixed fishery. These fishing grounds are also harvested by gillnet. Tables 4.3 and 4.4 show that 20–25% of the landings of ling are taken by gillnet, whereas insignificant proportions of the landings of tusk are taken by this fishing method. This may be due to a behavioural difference between tusk and ling. Tusk is regarded as a sluggish and stationary fish not known to undertake long migrations. The probability of tusk encountering a gillnet may therefore be low, whereas the bait odour released from baited longlines may trigger and attract tusk from a large area. The mobility of ling will bring this species into contact with both gillnets and longlines.

The groundfish stocks of cod and haddock in the Barents Sea were overexploited and at their lowest levels in the mid 1980s. These stocks have now recovered, and annual catches are increasing. Stock assessment work on tusk and ling is based on catch per unit effort data from the commercial longline fishery. The findings indicate a pronounced decrease in these stocks over the last two decades. The low catch quotas for cod and haddock in the late 1980s caused an increased fishing effort from the longline fleet on the tusk and ling stocks which are the only alternative for these vessels.

Most of the groundfish resources in the mid and northwest Atlantic are overexploited and are now at very low stock levels. The mean annual landings for the five-year period given in Tables 4.3, 4.4 and 4.5 show lower total landings in Norway than in Iceland and Canada. These figures reflect the fact that Norway has been through a period of significant restrictions, with low groundfish quotas. The Norwegian quotas have recently been increased whereas the groundfish stocks in Icelandic and Canadian waters have not recovered yet and the catch quotas for these areas are presently at very low levels.

4.3 Groundfish in the north eastern Pacific

Longlines, trawls and pots are the main gear types used to harvest the groundfish resources in the north eastern Pacific. The main demersal species in this area are cod, halibut and sablefish, of which cod is predominant in terms of quantity landed. Halibut and sablefish are almost exclusively caught by longlines, whereas cod is harvested mainly by trawlers (Table 4.6).

Table 4.6 Mean annual catches (tonnes) of groundfish, 1989–93, in north eastern Pacific by major gear types. The percentage of the landings taken by different gear types is given in brackets.

Species	Longline	Pot	Trawl	Other	Total
Pacific cod	70 420 (27)	12 640 (5)	173 420 (68)	20 (0)	256 500
Sablefish	22 184 (88)	4 (0)	3 104 (12)	–	25 292
Pacific halibut	27 427 (90)	–	–	2 910 (10)	30 337
Total	120 031 (38)	12 644 (4)	176 524 (57)	2 930 (1)	312 129

Sources: Alaska Fisheries Science Center, National Marine Fisheries Service, USA; Report of Assessment and Research Activities, 1993, International Pacific Halibut Commission.

Longline is the traditional gear used to harvest halibut, and today longlining is the only legal fishing method to be used in the directed fishery for halibut. After the Second World War, a foreign trawler fleet (mainly vessels from Japan and Russia) expanded and started a fishery in the north eastern Pacific. These trawlers caught large quantities of halibut, and their fishing effort represented a significant escalation of the exploitation of the halibut stock. Trawling was therefore banned to prevent foreign vessels from exploiting the halibut resources in the north eastern Pacific. This legislation has not been lifted, and was already in practice in the 1970s when the domestic fleet of trawlers started to operate on the fishing grounds inhabited by halibut. However, the bycatch of halibut in the domestic trawl fishery targeting Pacific cod is substantial. In fact, the halibut bycatch in this fishery is much higher than the directed longline fishery for halibut. Also in the longline fishery for Pacific cod there is a substantial bycatch of halibut.

Halibut caught in all fisheries except the halibut-directed longline fishery have to be discarded. The fate and survival of discarded fish are recognised as a serious problem among halibut fishermen. Halibut that are caught by trawl or longline, brought on board the vessel and then released from the gear and discarded have been subjected to severe stress and injuries. This controversial issue has revealed the need for research that would determine the fate of halibut that are discarded from trawlers and longliners. Such work is now being conducted by research institutes in several countries. Some preliminary results have been reported showing that mortality rates are dependent on how the fish are removed from the hooks, and for trawling on towing duration, catch volume and time spent on deck prior to release. With careful handling, high survival rates were reported – particularly for halibut caught by longlines.

The sablefish resource in the north eastern Pacific has been harvested since the early 1900s, but the annual catches were relatively small (averaging 1500 t in 1930–50) until the Japanese longline fleet expanded into the Gulf of Alaska. The catches rapidly escalated during the 1960s until the record catch from the Gulf of Alaska reached 35 700 t in 1972. They have averaged about 25 000 t in recent years (Table 4.6). Like the catches of halibut, the sablefish catches are nearly exclusively landed by longliners because longline is favoured for historical and political reasons. Halibut longliners historically also fished for sablefish, and this is the basic reason why longline is favoured in the fishery for sablefish. Trawlers and pot boats also capture sablefish, but their sablefish catch is restricted by regulations. The sablefish catch of trawlers is restricted to a percentage (4–15% depending on the area) of their weekly total catch. Pot boats can target sablefish in the Bering Sea and Aleutian Islands region, but not in the Gulf of Alaska where sablefish is currently most abundant.

Pacific cod was harvested by the Japanese longline fleet during the early 1960s, and cod became an important bycatch species in the Japanese trawl fishery for walleye pollock that expanded in the mid 1960s. In the early and mid 1980s, respectively, the US domestic trawl and longline fisheries expanded in the eastern Bering Sea and the Aleutian Islands region, and since 1991 the entire landings of cod have been taken by the domestic fisheries. The total catches of cod were below 100 000 t prior to 1980. Catches in the 1980s increased markedly because of increases in stock size as well as the fishing effort of the new US fleet. The total catches of cod in the eastern Bering Sea, Aleutian Islands region and Gulf of Alaska have averaged 256 500 t in 1989–93 (Table 4.6), and reached a historic high of 295 000 t in 1991 with the great majority originating from the eastern Bering Sea (208 300 t). Currently the Pacific cod stock is fished using trawl, longline, pot and jig gear, and future catches will be allocated between these fishing methods according to recent historic averages which means that the great majority of the total allowable catch will be taken by trawl and longline (see Table 4.6).

Assessment work on the halibut and sablefish stocks is based on commercial catch data and longline surveys to obtain abundance estimates. Over the last two decades the exploitable biomass of the halibut stock increased until the late 1980s, but since then has been decreasing. This is mainly due to a decrease in the recruitment which means that the stock will continue its annual decline for the next several years. Evidence of declining sablefish stock abundance led to significant fishery restrictions from 1977 to 1985. After 1979 the sablefish population increased significantly, with major increases in 1982 and 1985 which have been attributed to the growth of the strong 1977 and 1980 year classes. The population abundance indices indicate a slight decrease after 1985. Results from the eastern Bering Sea trawl surveys for cod indicate that the biomass increased from 1979 through 1983, and then remained relatively constant through 1989. The biomass estimates dropped by about 25% both in 1990 and 1991, but have since then shown an increasing trend with the 1994 estimate being the highest in the survey biomass time series.

4.4 Hake in the Atlantic

Important fisheries targeting hake are being conducted in the north eastern Atlantic and along the southern coasts of Africa and South America. Different species of hake are found in these areas, but they all belong to the same genus (*Merluccius*). The fishing method used by longline vessels targeting hake is based on the Spanish method of setting monofilament longlines attached to a safety line (see Section 3.2).

The fishery for hake in the north eastern Atlantic has traditionally been conducted by vessels from Spain, Portugal and France, with Spanish vessels taking more than 50% of the total international landings in this region. Longline and trawl are the dominant gear types in the hake-directed fishery (Table 4.7). The hake resources in the north eastern Atlantic are divided into a southern and a northern stock of which the latter is more important in terms of quantity landed. The total international landings were 10 900 and 52 150 t, respectively for the two stocks. The landings and biomass of both stocks have been decreasing sharply since the mid 1980s, and have now reached a historic low, probably as a result of overexploitation.

Table 4.7 Annual landings (tonnes) of hake by Spanish vessels in the north eastern Atlantic by major gear types. The percentage of the landings taken by different gear types is given in brackets. Means for 1989–93 are given for the southern stock, and data for 1993 (the only available data) are given for the northern stock.

Stock	Longline	Gillnet	Trawl	Total
Southern	2 220 (28)	2 040 (26)	3 620 (46)	7 880
Northern	14 814 (57)	–	11 280 (43)	26 094

Source: Report of the Working Group on the assessment of southern shelf demersal stocks. International Council for the Exploration of the Sea, C.M. 1995/Assess:6.

Two species of hake are targeted in the south eastern Atlantic (i.e. along the Namibian coast and the west and south coast of South Africa): the shallow water hake, occurring from the coast out to about 300 m depth, overlapped by the deep-water species living at depths from about 150 m to more than 600 m. The shallow-water hake is most valuable and is preferred for its texture. The hake resources in the south eastern Atlantic have traditionally been harvested by trawlers. The recent introduction of demersal longlining in this region has been, and is presently, a controversial issue in the fisheries.

The first demersal longline permits were issued in South Africa in 1983. Although initially intended for hake, longlining in South Africa was moved on to kingklip which was more abundant. The longline effort was increased as additional permits were given, and the total kingklip landings increased significantly in the mid 1980s. As the longline catches of klingklip began to drop sharply from a peak of 8684 t in 1986, the longliners again began to target more on hake. The longline catches of hake peaked in 1988 at 5514 t, and then dropped sharply. Demersal longlining was banned at the end of 1990, and at present the only legal longlining allowed in South Africa is the directed fishery for shark. Longlining for hake along the Namibian coastline has been practised since 1983, but the present policy is to eliminate longline in favour of bottom trawl, which is used almost exclusively to harvest the hake resources.

Longlining is regarded as a more size-selective fishing method than trawling, and concern has been raised that the removal of adult spawning hake and kingklip by the longline fleet may have had a strong impact on the spawning stock and recruitment to both the trawl and longline catches. However, historical catch records clearly show that the hake resources were overexploited by the trawler fleet in the 1970s when the mean annual catch off the coast of South Africa averaged 196 000 t. Assessment data on the kingklip resource also suggest that this stock was overexploited by the trawler fleet prior to the introduction of longlining, and that the longline effort only accelerated the decline.

The size selectivity of longlines allows the hake to grow to larger sizes before they are harvested compared with trawl (Fig. 4.1). Biological models therefore show that when the spawning stock is maintained at a certain sustainable level, harvesting by longlines would give a greater catch yield than would trawling. Although these results of modelling biological effects suggest that longlining is preferable to trawling, there are still many questions about the effects of longlining that need to be addressed. A longline investigation is currently being conducted in South Africa to provide the necessary background information on deciding if the ban on demersal longlining for hake should be lifted. The scientists working on this project, and Namibian scientists also, recommend that longlining should be preferred over trawling for resource conservation reasons.

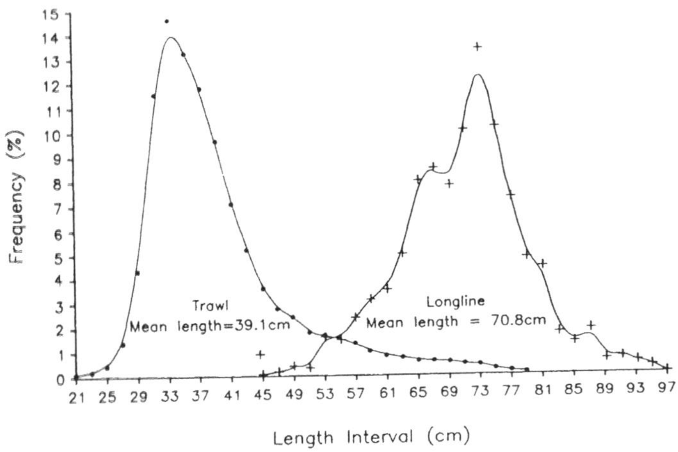

Fig. 4.1 Comparison of the length–frequency distributions of hake caught in the South African demersal trawl and longline fisheries. (Redrawn after Japp, 1993.)

In the south western Atlantic (i.e. along the coast of Argentina), longline fishing activities started in the late 1980s. Because of the low price of hake in this region, hake is mainly caught as bycatch in the fishery targeting kingklip. The most valuable fish caught in this area, however, is the Patagonian toothfish (US$6.70 per kilogram headed and gutted). In recent years, only two or three longliners have therefore been targeting kingklip and hake, while the rest of the longline fleet has been engaged in the directed fishery for Patagonian toothfish.

Chapter 5
Gear Parameters and Catching Efficiency

5.1 Research methods

In attempts to improve the catching efficiency and selectivity of longline gear, several research methods are applied. The scientific methods used in gear improvement programmes can be classified into two categories: behaviour studies and fishing trials, each of which has two levels. Behaviour studies can be carried out in the laboratory or in the field, and fishing trials can be conducted as experimental fishing or full-scale trials.

Behaviour studies in the laboratory with relevance to longline gear are conducted by holding fish in tanks and observing their response to different gear parameters. Several studies have been made in the laboratory, e.g. to identify feeding stimulants and to observe responses to different bait and hook types (see Fig. 5.1). The advantages of con-

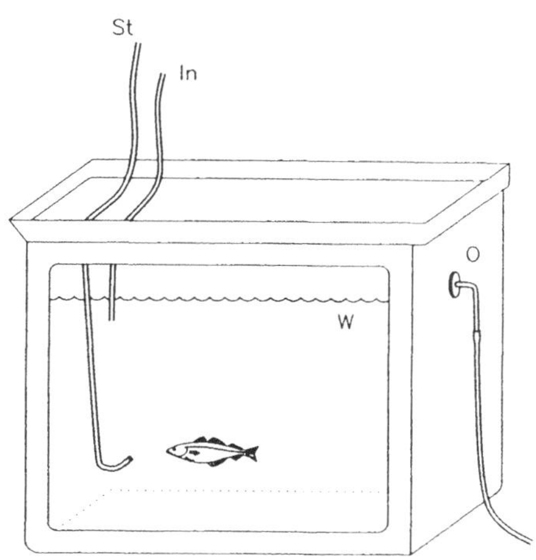

Fig. 5.1 Experimental tank for identifying feeding stimulants in fish. The stimulatory capacities of different chemical compounds are determined by recording the number of times a fish attacks the stimulant injection port. St, Stimulant injection tube; In, sea water inflow; O, sea water overflow; W, observation window.

ducting laboratory studies in a gear improvement programme are that detailed behavioural observations can be obtained and that the fish observed are under control in terms of species, size, feeding motivation, diet experience, etc. Thus the history of the fish under observation is known. This is important because the responses of fish to baited gear are affected by the size of the fish, its hunger level and previous diet experience, competition and other factors. The main problem of conducting behaviour studies in the laboratory is that the fish are not observed in their natural environment. The information obtained from these studies may therefore be affected by artefacts related to laboratory conditions, and should therefore be considered as indicative rather than conclusive with respect to expected responses in the field.

This problem is overcome in behaviour studies carried out in the field. In these studies an underwater television camera is used to observe fish responses to the gear (Fig. 5.2). Such studies have been made from the shore, from boats, and from oil platforms. The last of these methods ensures observations under weather-independent conditions at locations that normally have high fish densities. The efficiency of different hook and bait types, diurnal rhythms of activity toward baited hooks, and hooking behaviour of different species have been studied in the field by underwater cameras. Because visibility in water is limited, the observation range of underwater cameras is limited to a few metres and this method is therefore useful only for detailed behaviour observations of fish in the close vicinity of the gear. Tele-

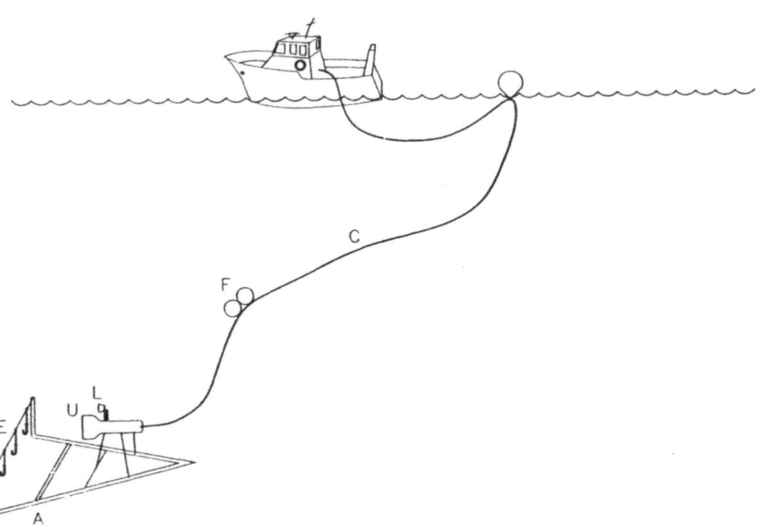

Fig. 5.2 Observation technique for fish behaviour study in the field. A, Aluminium frame; E, experimental longline; U, underwater camera; L, lamp; F, floats; and C, cable. (Redrawn after Sutterlin *et al.*, 1982.)

metry (the use of ultrasonic transmitters attached to animals to track their movements) has been applied for studying the responses of fish before they come into close contact with the gear. Tagging fish with acoustic transmitters has made it possible to study their capability to respond to baited hooks at distances of several hundred metres (see Section 2.3).

Experimental or comparative fishing trials are commonly conducted on board commercial longliners. In these studies the effect of different gear parameters on the catching efficiency and selectivity can be investigated in actual fishing. For example, the efficiency of a new hook design can be tested by rigging the longline with this hook type and with traditional hooks in alternating clusters of 50–100 hooks. The method has been used to test different types and sizes of hook and bait, different snood and mainline materials, and swivel-mounted snoods. Bait loss, the effect of fishing time and the importance of chemical stimuli in attracting fish have also been investigated using this type of experimental fishing.

When conducting comparative fishing trials, it is important to test only one gear parameter in each comparison while keeping all other parameters constant. For example, when testing the effects of hook size and bait size, this should be carried out in two separate trials. If, for example, an experimental longline that is both rigged with larger hooks and baited with larger baits is compared with a traditional longline and found to catch larger fish, no conclusion can be drawn as to whether this size-selective effect is caused by increased hook size or by the bait size. When designing and conducting comparative fishing trials, it is also important to record all types of variable that may be affected by the gear modification, e.g. catch rates of all species including target species, bycatch and trashfish, the size of the fish caught, bait loss and line tangles. Soak time, fishing depth and current speed and direction may also be of importance when interpreting the results obtained. The data must be recorded directly during the hauling of the gear, often a challenging task under the varying conditions during sea trials on board commercial vessels. The use of hand-held data field recorders has made the data collection easier and has also significantly rationalised the subsequent data analysis.

Comparative fishing trials provide strong indications of the effect of a gear modification, but have limited value in terms of giving the behavioural explanations for the observed differences, e.g. explaining why small baits catch more smaller fish than larger baits. The difference in size composition between catches taken with small and large baits is best determined by conducting comparative fishing trials, but behaviour studies have to be made to explain why small fish do not take large baits.

Full-scale and long-term fishing trials are the final test of the performance of a gear parameter. In a full-scale fishing trial the modified gear, e.g. with a new hook design or swivel-mounted snood, is deployed in commercial longlining and tested during a whole fishing season to reveal its catching properties under different conditions. This will provide conclusive evidence of the catching efficiency of the modified gear, but, like experimental fishing trials, the data obtained have limited explanatory value. A summary of the advantages and disadvantages of the different methods for investigating longline gear modifications is given in Table 5.1.

Table 5.1 The major advantages and disadvantages of four different scientific methods used in longline gear research. (Modified after Løkkeborg *et al.*, 1993.)

Method	Advantages	Disadvantages
Behaviour studies in the laboratory	Control of fish in terms of number, species, diet experience, feeding motivation Control of environmental conditions regarding temperature, light level, etc. Ideal observation conditions Individual fish can be identified and observed Low costs	Fish may adapt to laboratory conditions Fish do not feed on their natural diet Motivational state may differ from that of wild fish Conditioning to gear Difficult to establish bait odour gradients
Behaviour studies in the field	Representative observations of behaviour and activity of wild fish	No control of fish in terms of number, species, size, etc. Restricted observational area Time-consuming
Comparative fishing trials	Strong indications of the effect of a gear modification	Give limited explanations of the results High effort and costs
Full-scale and long-term fishing trials	Conclusive evidence of catching effectiveness	Give limited explanations of the results Convincing evidence from previous levels required Dependent on cooperation with fishermen

5.2 *Natural bait*

Longline fishing is based on stimulating the food-searching behaviour of fish through the odour released from the baits. Once the fish have been attracted to the longline by the bait odour, the baits serve to trigger the fish to attack and ingest the baited hooks. The number of fish that encounter a baited longline and ingest the baited hooks is determined by the stimulating capacity of the smell and taste stimuli generated by the baits, i.e. the concentration and composition of the chemical compounds released by the baits. The type and quality of the bait are therefore very important factors for the catching efficiency of longlines.

The fisherman has learned from experience that different bait types give different catching efficiencies for different species (see Section 6.2, Species selection). The type of bait chosen for a specific fishing operation is therefore determined by the target species (Table 5.2). In the fishery for demersal species in the north east Atlantic, for example, saithe is normally used as bait for wolffish, and squid or mackerel are used when targeting cod and haddock. In addition to the attractiveness of the smell and taste stimuli, the efficiency of a bait is determined by its physical strength and ability to remain on the hook throughout the soaking period. In the fishery for cod and haddock, mackerel has been proven to give better catch results than squid when pelagic or semi-pelagic longlines are used. Squid, however, withstand bait loss better than mackerel, and when fishing cod and haddock with bottom long-lines the fishermen bait their longlines with squid or a combination of squid and mackerel. The bait loss is more significant for hooks on the bottom, and squid may therefore be more efficient for bottom longlines than mackerel even though mackerel may be more attractive to cod and haddock.

Using two different types of bait on the same longline has in some fisheries resulted in improved catch rates, particularly in mixed-species fisheries. There might be two reasons why a combination of two bait types on the same longline gives better catch results than longlines baited with one of the two types only. As different bait types vary in attractiveness and ability to withstand bait loss, a longline baited with a combination of an attractive bait and a strong bait has the ability to attract a large proportion of fish from a wide area and also has a high proportion of hooks with intact baits throughout most of the soak time. In mixed-species fisheries the various target species may have a preference for different bait types. Using a combination of baits may therefore give higher total catch rates, although longlines baited with one bait type only may give a higher catch rate for one of the species. In the longline fishery for tusk and ling, a combination of mackerel and squid baits is commonly used. For longlines baited with mackerel and squid in a 4:1 ratio (every fifth hook baited with squid), it has been shown that the squid bait gave 100% higher catch rate for ling than the mackerel bait but only 9% more tusk, indicating a preference in ling for squid bait. Another fishing experiment showed that mackerel and squid bait at a 1:1 ratio caught 40% more tusk than longlines baited with mackerel only.

In addition to the type of bait used, the quality of the bait is of significant importance for the catching efficiency. The quality and quantity of odour released by a bait may be affected by the feeding history of the bait specimen. The mackerel feed heavily during the summer, but do not feed much in the winter. The protein and fat contents are

Table 5.2 The most common bait types used to target different species

Bait type	Target fish species												
	Atlantic cod	Pacific cod	Haddock	Tusk	Ling	Sablefish	Atlantic halibut	Pacific halibut	Greenland halibut	Wolffish	Hake	Atlantic salmon	Tuna
Mackerel	*		*	*	*				*				*
Squid/ octopus	*	*	*	*	*	*		*	*				*
Shrimp	*												
Herring		*				*		*					
Salmon								*					
Saithe							*			*			
Tusk							*						
Sardine											*		*
Sprat												*	
Saury													*
Scad													*
Milk fish													*

therefore different for mackerel caught in different seasons. Bait from mackerel caught during the summer season when the mackerel feed on copepods has given much higher catches of haddock and tusk than bait of mackerel caught in the late autumn (Fig. 5.3). Thus, bait of the same type but caught in different seasons or areas may give different catching efficiencies.

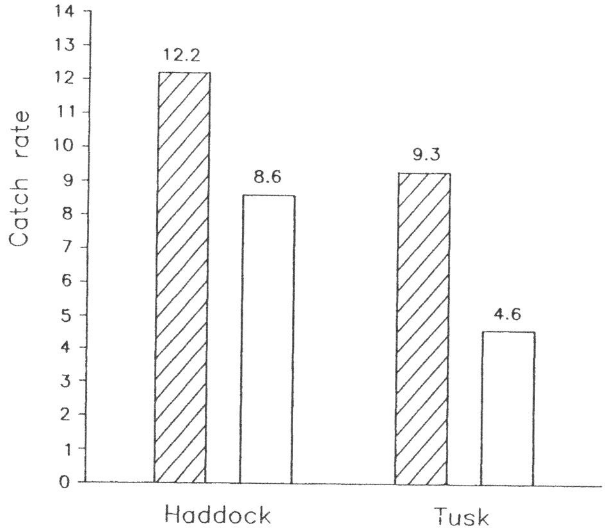

Fig. 5.3 Longline catch rates of haddock and tusk obtained with baits of mackerel caught in summer (hatched bars) and autumn (open bars).

Furthermore, proper handling and storage of the bait after it has been caught are also of importance for the bait quality and hence the catching efficiency. The decomposition process that changes the chemical composition of the bait may start very quickly and go fast once it has started. The bait must therefore be cooled and frozen as soon as possible after capture, and kept frozen until baiting. If the longlines are not set shortly after baiting, they should be kept in cold storage to preserve the bait quality.

The bait size is an important factor affecting the size of fish caught by longlines. This gear parameter is therefore described in more detail in Section 6.3 which covers size selection. Both the size (size selectivity) and number of fish (catching efficiency) caught are affected by the size of the bait. The bait cost represents a high proportion (about 20%) of the total cost of the longline operation, and using smaller baits therefore significantly reduces the expense. It has been shown that using baits of half the size normally used in the longline fishery for tusk and ling did not cause a reduction in the catch rate for tusk and even gave a 25% catch increase for ling. The most pronounced effect of bait size on catching efficiency has been demonstrated for haddock with semi-pelagic longlines. A 10-g mackerel bait was in this fishery shown to catch more than twice as many haddock as a 30-g mackerel bait (Fig. 5.4). Baits in midwater are more readily seen than baits on the seabed, and the effect of bait size is therefore likely to be more pronounced for pelagic or semipelagic longlines than for bottom-set gear.

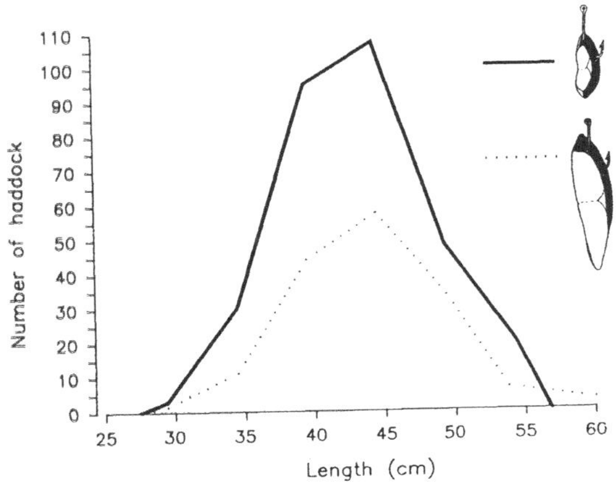

Fig. 5.4 Comparison of the length–frequency distributions of haddock caught on mackerel baits of two different sizes, 10 g (solid line) and 30 g (dotted line), respectively. (Modified after Johannessen, 1983.)

5.3 *Artificial bait*

It has been the dream of many a fisherman to have a 'super bait' with extremely high catching power for the desired target species, and that this bait could also be used over and over again with no need to rebait the hook. Unfortunately, such a bait is still nothing but a dream. Work aimed at developing artificial longline baits has been conducted in several countries for many years, but none of the artificial baits tested so far seems to be effective enough to meet the requirements in terms of catchability and cost of replacing the natural baits used in commercial longlining. There are several reasons for developing an artificial longline bait. First, large quantities of bait are used annually – resources which could instead be used for human consumption. During a 6- to 7-week fishing trip a Norwegian autoliner uses 25–30 t of bait. There are no good quantitative figures for global bait use, but to illustrate the magnitude the annual bait consumption in Norway alone is in the order of 15 000–20 000 t. Several fish stocks are overexploited. With increasing demand for marine resources for human consumption, high quality bait resources at an affordable price to the longline fleet may become less available in the near future. Secondly, the selectivity in longlining may be improved with an artificial bait, as important factors affecting species and size selectivity, such as taste, size and shape of the bait, can be varied within larger ranges. Results obtained so far indicate improved species and size selectivity for some of the artificial baits tested in commercial longlining. Thirdly, an artificial bait can be cut in uniform pieces, and thereby make mechanised baiting easier and more precise. Thus, more simple and cheaper baiting machines may be designed for smaller vessels. Finally, an artificial bait made from synthetic chemicals could be stored at room temperature, and additional freezing costs would therefore be saved.

Two different general approaches have been made in the attempts to develop artificial longline baits: that using baits based on extracts or synthetic chemicals as stimulants (synthetic bait) or baits based on natural surplus products from the fishing industry as inexpensive stimulants (restructured bait; see Fig. 5.5). In both cases the artificial baits consists of three main components. A *reinforcement* is used to give the bait physical strength to stay on the hook and withstand bait loss. Baits based on surplus products are often reinforced by means of a stocking or bag made from fine-meshed fabric. For baits with extracts or synthetic chemicals as stimulants, the reinforcement also serves as a filling material that holds the stimulants, and foam rubber materials such as polyurethane have been used. To trigger the fish to search for and ingest the baited hook, the bait must release effective *feeding attractants and stimulants.* The third component of an artificial bait is a *binder*, which is used to regulate the release of attractants over an adequate period of time. Without a binder the bait will lose its stimulating properties shortly after the longline has been set because the attractants are washed out too quickly.

The catching effectiveness of an artificial bait is determined by the properties of these three components. The physical strength of the reinforcement must prevent bait loss due to birds, bait scavengers and fish attacking the bait without being hooked, and thereby ensure that most of the baits stay on the hook throughout the soaking period. A complete mixture of feeding attractants and stimulants is necessary in order to effectively attract and stimulate the fish to ingest the baited hook. To maintain the stimulatory capacity of the bait for an extended period of time, the binder must have chemical properties that give a certain release rate of attractants over most of the soak time. The attractant release rate can be matched to that of natural baits in accordance with the type and amount of binder material to be used in the bait mixture (see Fig. 2.5). However, with artificial baits there is also the possibility of regulating the attractant release profile in order to maximise the catching efficiency of specific longline fisheries. In addition, the texture and consistency of the bait may affect the catchability for some species.

Thus in the development of an artificial bait to replace traditional bait types, reinforcement, stimulants and binder materials that give strength, stimulatory capacities and release rates similar to or better

(a)

(b)

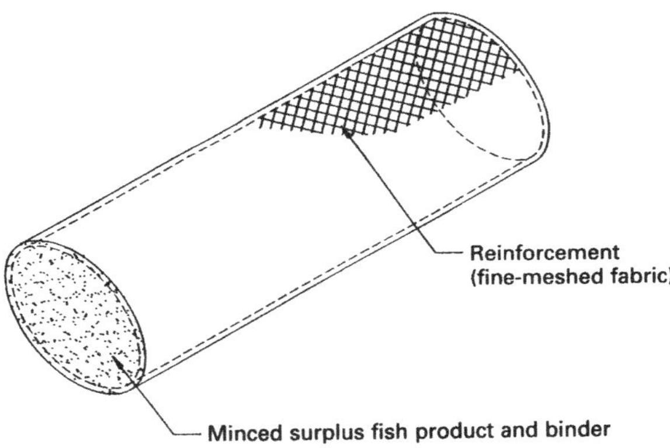

Fig. 5.5 Two different approaches to the challenge of developing an artificial longline bait. (a) Bait based on synthetic chemicals identified as feeding stimulants in shrimp (right); and (b) bait based on surplus products from the fishing industry.

than natural baits are required. Finding a reinforcement that fulfils the requirements seems to be the easiest task. The reinforcement should not be too strong as old baits have to be cleaned from the hooks before rebaiting. The reinforcement should also be biodegradable.

The most critical task in the development of an artificial bait based on extracts or synthetic chemicals seems to be the composition of the stimulant mixture. Many studies have been conducted to identify the feeding stimulants for different species of fish and crustaceans. These studies have revealed that a mixture of several substances rather than a single dominant substance is required to yield synthetic stimulants with the capability of effectively triggering the feeding behaviour of fish. Therefore, the complete mixture of feeding substances has to be identified in order to make the best composition of synthetic chemicals for an effective artificial bait. Without all individual substances identified, an artificial bait may lack critical components which may explain the low catch rates obtained for some of the baits developed so far. In addition, feeding stimulants are species specific, i.e. the major feeding stimulants are different for different species. Furthermore, the food preference of a certain species may change between seasons and is affected by previous diet experiences. A bait developed for a particular longline fishery may therefore be effective for only one or a few species during a specific season. All these factors together make the task of identifying feeding attractants and stimulants for an effective artificial bait very complicated. On the other hand, because different species respond to different mixtures of stimulants, artificial bait development gives unique potential for improving the species selectivity of longlining.

When developing an artificial bait based on surplus products, one is not faced with the problem of composing a mixture that contains all individual feeding substances as the fish tissue used should contain the entire attractant and stimulant milieu. However, there remains the problem of finding surplus fish resources that have stimulatory capacities similar to those of natural baits. An interesting approach is to include a digestive enzyme in this bait type in order to increase the concentration of attractants. As amino acids constitute the main group of feeding attractants and stimulants, an enzyme that disintegrates the proteins of fish tissue into amino acids may elevate the attractant levels.

A binder that gives an appropriate release rate of attractants is also critical for the catching efficiency of an artificial bait. The release rate of natural baits has been shown to be initially high with a rapid decrease during the first hours of soaking in seawater (see Fig. 2.5). This may be the optimal release rate when targeting species that have a clear and relatively short period of high feeding activity and the baited longlines are set at the beginning of this period. To catch fish that do not exhibit a pronounced rhythm of feeding activity but are active over longer periods, baits having a constant release rate which lasts in accordance with the soak time of the gear may be more effective. Methods have been evolved to determine the release rates of attractants from baits, and binders of different chemical properties can be tested in different concentrations to obtain the desired attractant release characteristics.

Baits with a long-lasting taste component may be achieved by applying encapsulation technology where stimulants are placed in micro-capsules from which they are released at a slow rate.

Although efforts on artificial bait development have not yet produced a bait that meets the requirements for replacing natural baits, some very promising results have been obtained. In Norway comprehensive studies have been conducted to develop baits based on both synthetic chemicals and natural fish resources, partly as collaborative studies with research institutes in other countries. In cooperation with the University of Florida a synthetic shrimp-flavoured bait was developed. This bait gave almost the same catch rates as natural shrimp bait, and did improve the size selectivity as the proportion of undersized cod was lower in the catches taken by this bait type than in the catches taken by natural bait. However, this artificial bait proved to be effective only during a short, seasonal fishery where shrimp is used as bait. It compared poorly with natural bait in other seasons, and has therefore not been applied in commercial longlining. A similar shrimp-flavoured artificial bait has been developed for recreational hook-and-line fishing, and is commercially available.

Countries with large fishing fleets and fishing industries have high quantities of marine resources that are not used for human consumption. Resources such as trashfish and surplus products from fish processing plants (offal) may be applied as stimulants for an alternative longline bait. Bait development work based on this idea has been conducted in several countries (e.g. Canada, Norway, Spain and the USA). In recent Norwegian investigations, promising catch results have been obtained with a bait based on minced surplus products from fillet production (mackerel and herring) with alginate as binder agent. This mixture of minced fish and binder is filled into long stockings made from viscose fabric, using the same technique as for sausage production. Preliminary fishing trials have been conducted for the most important species (cod, haddock, tusk, ling and wolffish) in Norwegian longlining. The catch results obtained are promising – particularly the catch rates for haddock, which were several times higher than with traditional mackerel bait. Encouraged by these interesting results, an increased research effort has been initiated to achieve the objective of developing alternative longline baits that fulfil the competitive requirements in terms of catching efficiency and cost. If an effective restructured bait based mainly on fish offal could be developed, future longline baits could to a large extent be produced from the fish offal of the longline industry. As an example, an autoline vessel normally lands 100–150 t of dressed fish from a 4- to 6-week trip. During the same trip 30–50 t of offal have been discarded at sea, resources that, if recycled as bait, would more than satisfy the bait demand (normally around 20–25 t during the same time period).

In addition to artificial baits based on fish attraction by chemical stimuli, several attempts have been made to investigate the effect of light stimuli on longline hooks. The basic technology applied has been the use of light crystal materials that after being charged in daylight or artificial light emit a weak light for several hours (a principle similar to that used in emergency exit signs in cinemas, hotels and the like). The light crystals have been embedded in discs of foam rubber or plastic foil and applied to the hooks in addition to normal bait. In some tuna longline fisheries, luminescent material is applied at the connection between the snood and the hook. For tuna, being a visual feeder, the use of these light-stimulating devices may have a positive effect. To the authors' knowledge, light stimuli are not used in any other longline fisheries.

5.4 The hook

Until recently the traditional J-shaped hook was the commonly used longline hook. This hook type entirely dominated the longline fisheries for groundfish until the mid 1980s. Fishing experiments with new hook designs have shown that considerable increases in catch rates can be obtained by using improved hook types. The most pronounced improvement with new hook types has been obtained in the longline fishery for halibut. Research conducted by the International Pacific Halibut Commission has shown that, on average, the catch rate of the Circle hook may be 2.2 times higher than that of the J-hook. Likewise, the EZ-baiter hook, which has an intermediate shape between the Circle hook and the J-hook, has given significantly improved catch rates for different codfishes. After some early scepticism among fishermen against the new hook designs there has been a widespread change from traditional hooks to new, different hook designs (Circle, EZ-baiter, Wide Gap) that are now widely used in various longline fisheries. As an example of this development, Fig. 5.6 shows the marked and rapid change from the J-hook to the EZ-baiter hook in the Norwegian longline fishery.

Several factors may affect the catching efficiency of a longline hook, but the shape of the hook is most important. The new hook types developed for use in fisheries for different species have one feature in common, namely that the point of the hook is bent towards the eye or the shank (Fig. 5.7). A hook with this shape has the point towards the line of pull, which ensures that the tension placed on the snood is more effectively transferred to the point of the hook (Fig. 5.8). For the J-hook there is an angle between the line of pull from the snood and the force generated on the hook point against the mouth tissue. The penetrating force applied by the point of the hook will therefore be somewhat less than the actual pull force of the snood, with a correspondingly lower hooking probability.

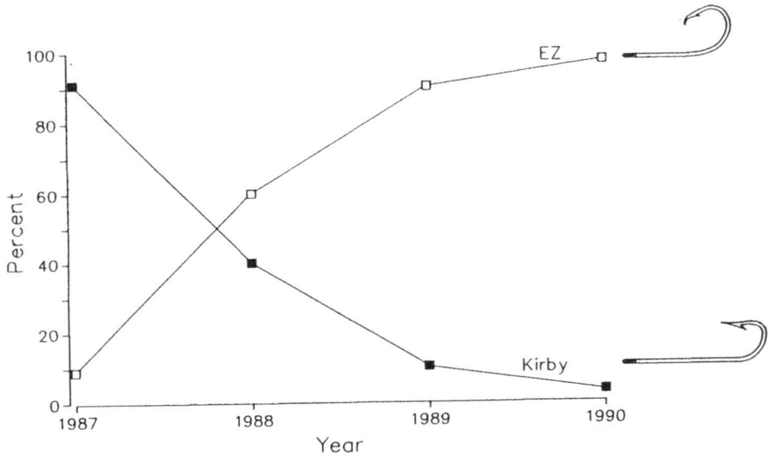

Fig. 5.6 The marked shift from the traditional J- (Kirby) hook to the EZ-baiter hook in the Norwegian longline fishery, given as proportion of hooks sold. (Source: sales statistics, O. Mustad & Son A.S.) (Redrawn after Løkkeborg *et al.*, 1993.)

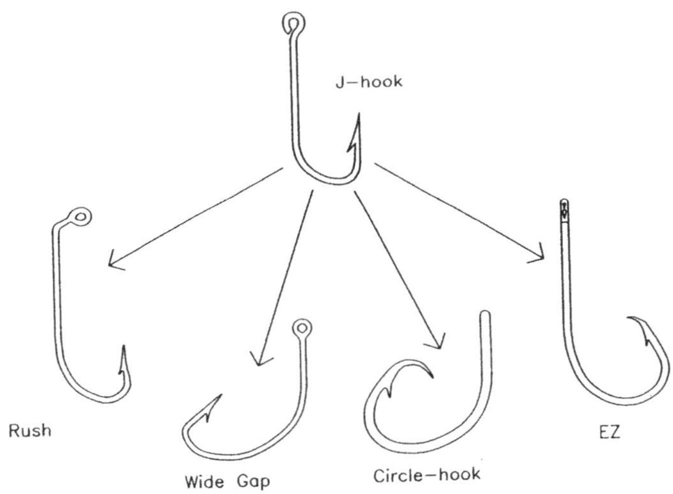

Fig. 5.7 Development of new hook designs showing the traditional J-hook and the more effective hooks with the point bent towards the eye or the shank.

Another factor that may explain the superiority of the new hook designs is their ability to prevent hooked fish from escaping. The gape is more narrow when the point is bent towards the eye or shank, and this may mean that hooked fish are retained more effectively when they are struggling to escape. This is supported by the fact that fishermen have experienced the highest catch increases with the improved hook

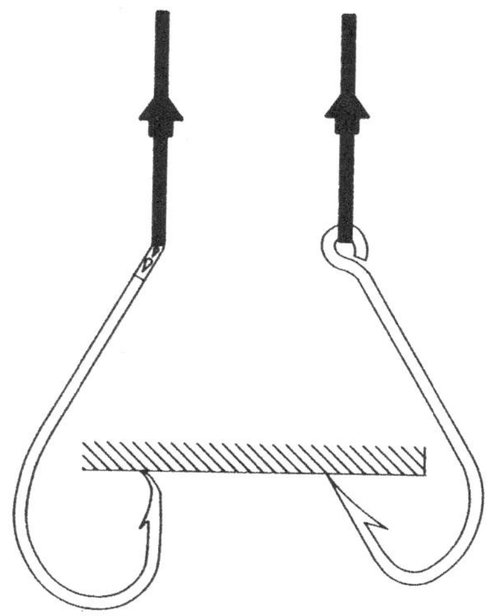

Fig. 5.8 The line of pull for EZ-hook and J-hook showing that the hook point of an EZ-hook is directed along the tension made on the snood, whereas the point of a J-hook makes an angle with the snood.

designs under bad weather conditions, when it is of particular importance that the hook has a good hold on the fish.

The size of the hook also affects the catching efficiency in longlining. Smaller hooks generally give higher catch rates than larger hooks. The higher hooking probability of smaller hooks may have two explanations. When biting a bait on a small hook, a fish more easily also takes the hook into its mouth and gets hooked. Secondly, a stronger force is needed to pull a hook made from a thick thread into the mouth tissue. Smaller hooks are thinner, and therefore penetrate the flesh more easily. If the hook is too thin, however, it may be broken by large fish when the line is retrieved. How hook size may affect the size of fish caught is discussed in Section 6.3.

The longline hooks discussed so far are all single-point hooks. However, in sport fishing treble and double hooks with three and two points are commonly used, presumably to increase the catching efficiency. For this reason trials were made with double hooks in the longline fishery to test their catching performance in comparison with traditional single hooks. In the fishery for ling and tusk, the double hooks did indeed produce better catch rates than the traditional single hooks. However, this did not lead to an implementation of double hooks in commercial longlining. This illustrates an important feature of longline gear improvement: even though a new gear development is found to be beneficial from a pure catch performance aspect, the effect of changing gear configuration on the effective handling of the gear must also be borne in mind. In the case of the double hook, although it caught more fish, the handling of this hook was far more complicated and time consuming than the handling of standard hooks and thus the overall evaluation of the double hook was not positive from the fisherman's point of view. The double hook gave higher catches per unit of effort, but as significantly fewer hooks could be handled per unit time the relative catch increase was counterbalanced by the lower effort (daily number of hooks), resulting in a similar or lower daily catch than that obtained by traditional hooks. This also applies to different hooks and other changes of the longline gear that give higher catch rates *per se* but give rise to poor handling properties that outbalance the isolated catch rate benefits from reduced fishing effort so that the net gain may be rather insignificant.

5.5 The snood and snood attachment

The catch rates of longlines may be affected by the characteristics of the snood, i.e. the material, the thickness, the length and the attachment (how it is linked to the mainline). Investigations have shown that monofilament snoods give higher catch rates for cod and haddock (a 10–20% improvement) as compared with multifilament snoods, and that thinner snoods tend to give better catch rates than thicker ones. However, there is a compromise between finer materials and thinner dimensions that improve the catching possibility and the risk of breaking the snood so that the fish are lost. Thus the material and snood thickness have to be judged carefully according to target species and type of fishery. The lower visibility of monofilament snoods is commonly used to explain why these snoods give better catches than those made of multifilament. However, this has never been documented, so the real reason for the difference in catching power between multi- and monofilament material is still unclear.

Comparative fishing trials have shown that catch rates are generally improved with increasing snood length. For practical reasons snood length is normally limited to less than half the hook spacing as longer snoods tend to tangle more easily. The reason why longer snoods give higher catch rates is not fully understood. However, when the fish is hauled through the water it tends to rotate, both around its own axis and around the mainline. This often results in considerable twisting and shortening of the snood (Fig. 5.9) so that its flexibility is significantly reduced and hence the possibility of the fish breaking loose and getting lost is increased. A longer snood takes more twisting before it is shortened to a critical point with respect to loss of fish, which may explain the better catching performance of longer snoods.

Twisting of snoods is a problem with traditional rigging of longlines, where the snoods are attached by tying them directly to the mainline. With the introduction of monofilament lines, traditional snood attachment by tying was difficult because of the stiff and slippery

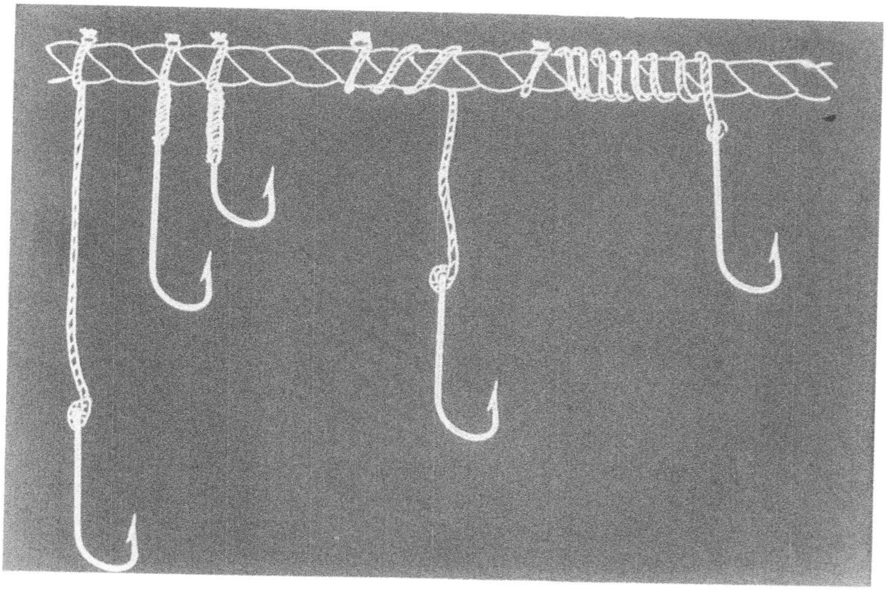

Fig. 5.9 Twisting, tangling and corresponding shortening of snoods caused by fish rotating during hauling.

character of the monofilament material. The common solution was to use swivels on the mainline to which the snoods were attached. The higher effectiveness of monofilament gear can partly be explained by this feature, as the swivel prevents twisting and tangling of the snood. In order to improve the catching capability of multifilament longlines, trials with swivel-connected snoods were done by Norwegian researchers in the mid 1980s. After some initial problems with swivels that were too weak, functional multifilament swivel lines were developed (Fig. 5.10). Through a series of comparative fishing trials it has been proven that the swivel attachment improves the catch rates by about 15%, depending on the type of fishery, the target species and weather conditions during hauling of the gear. This gain is explained by reduced loss of fish during hauling, as the swivel prevents twisting of the snoods so that their flexibility is maintained. During the late 1980s swivel gear replaced traditional multifilament gear in most of the Norwegian longline fishery, and is now also being used in several other parts of the world. When using swivel-connected snoods, fishermen also found that the work of de-twisting snoods was significantly reduced, and reports from autoline vessels told that they could reduce by one person the team that maintained the gear. The swivel also provides the possibility of using monofilament snoods with multifilament mainlines, and this has become common practice in the Norwegian coastal fishery for cod and haddock.

Fig. 5.10 Different multifilament swivel lines: Quick Snap (O. Mustad) (top), Fiskevegn (middle) and Nor-Lanka (bottom).

5.6 Hook spacing

The variations in hook spacing of longline gear are considerable. The hook intervals commonly used in a particular fishery have for the most part been developed by trial and error. Variations in hook spacing are often related to the fish density of the target species. In longlining for target species that have a dense distribution, such as spiny dogfish, cod and haddock, the hook spacing is close (1–3 m) whereas in longline fisheries for tuna and halibut, which are less densely distributed, the hook spacing is much longer. With most mechanised longline systems, however, the flexibility of varying the hook spacing is limited. In the most commonly used systems the hooks are stored on rack magazines. Because the mainline hangs as bights, the hook spacing is usually less than 1.5 m to avoid tangling of the bights.

Hook spacing may also vary within a fishery for the same species. In the Pacific halibut fishery, the interval between hooks has varied from 3 to 12 m. In longlining for tuna, hook spacing from 24 m to as much as 180 m has been used. Because the stock assessment work on Pacific halibut is based on longline catch data, a comprehensive study has been conducted by the International Pacific Halibut Commission to investigate the effect of hook spacing on catching efficiency. From this study and similar work conducted in other fisheries, it can be concluded that the catch per hook increases with increased hook spacing. The difference in the catch per hook at different hook spacings is related to competition between adjacent hooks. With close intervals between the hooks, several hooks compete for the capture of the same fish because the hooks are more densely spaced than the fish that encounter the gear, i.e. several hooks fall within the feeding range of each fish. The catch per hook therefore increases with increasing hook spacing until the hooks are so widely spaced that each fish encounters no more than one hook.

On the other hand, the catch per line decreases with increased hook spacing. With fewer hooks per line, i.e. longer hook spacing, the concentration of attractants released from the baits is lower and the line attracts fish from a smaller area. Furthermore, if the hooks are too far apart, there may be sections along the line where the number of fish encountering the gear is higher than the number of hooks available. In this situation, several fish compete for the same hook, and the gear becomes 'saturated'. As indicated by the above mentioned factors, it is difficult to obtain an optimal hook spacing under all conditions. In practical terms the hook spacing is therefore (based on experience) a compromise between the distribution pattern of the target species, the cost of the bait and the baiting method (manual, mechanical). Fishermen attempting to obtain a maximum catch with the minimum of effort should carefully consider these factors to attain the optimum spacing of their hooks.

5.7 The mainline

Monofilament longlines have been shown to be far more effective than multifilament gear in different longline fisheries, and differences in catching efficiency as high as 300% have been documented. As mentioned for monofilament snoods, the reasons for this are not completely clear but lower visibility of mainline and snoods, the 'fineness' (thin dimensions) of the gear and the mode of setting (pelagic, semipelagic) have commonly been used to explain the superiority of the monofilament longlines. However, because of the limited strength of the material and the relatively low resistance to chafing, the use of monofilament is restricted to relatively shallow water and to semipelagic or pelagic fishing. In addition, the handling of monofilament gear is more difficult and there is still no good solution for mechanised operation of monofilament lines, which also limits the use of this otherwise catch-effective longline gear.

As for monofilament lines, different experiments have indicated that the catching efficiency increases with thinner multifilament mainlines (reduced diameter). However, as stated above, there must always be a compromise between the desire for high catchability and the increased risk of losing the thinner lines. Questions are often raised about the possible repelling effects resulting from the use of strong-smelling impregnating agents (e.g. coal tar, Stenoline etc.) on new multifilament mainlines. The general experience is that new lines usually give significantly better catch rates than used gear, and no evidence has been produced on any repelling effects caused by the smell from impregnating agents. Questions have also been raised about the possible fish-repelling effects of the rotting smell created by fat and juices from bait that are absorbed by the multifilament gear ('sour lines'). The separate effect of 'sour mainlines' has never been investigated properly, and the reduced catch rates obtained with sour gear might be caused by other factors: the fact that the line has become sour indicates that it has been used for some time, which again implies that many of the hooks and snoods might not be in the best shape, thus giving reduced hooking probability and increased risk of losing fish during hauling. The sour smell might also camouflage the fresh smell from the bait and actually confuse the fish when trying to locate the baited hooks so that the overall catching efficiency is reduced. This effect has never been observed with longlines, but underwater observations of lobsters approaching a pot have shown that the animal may spend time investigating the smell traces on the pot line which thereby distract it from the desired direct movement towards the pot.

5.8 *Bait loss and other factors affecting the catching efficiency of longlines*

In the preceding sections the major longline gear parameters were considered with respect to catching efficiency. To summarise and to include other non-gear factors that have a significant impact on the overall catching power of the gear, it would be useful to follow the hook through the capture cycle from when it leaves the vessel until it is retrieved. This cycle may be divided into four main stages: setting, sinking, soaking and retrieval.

Bait loss

The risk of losing the bait from the hook varies considerably in longlining, and there are several sources of bait loss. First of all, bait loss can be purely mechanical during setting of the gear, caused by a combination of weak (soft) bait and the forces applied when the hook accelerates out of the tub or baiting machine, or when it hits the line chute or the sea surface. During setting, the baited hooks are also exposed to sea birds from the time when the hooks leave the vessel until they sink out of the birds' feeding range (1–2 m depth). This means that birds that normally stay around longline vessels in great numbers (e.g. fulmars, kittiwakes) have access to the baited hooks and might steal bait for up to 10–100 m behind the vessel. This bird feeding zone increases with larger vessels as the line is paid out higher above the sea surface with increasing vessel size (Fig. 5.11). The use of mechanised baiting also often increases the bird feeding zone as the line is normally set with a higher tension than when tub gear is used. Bait loss due to bird predation had been regarded as a minor problem until it was documented that birds could cause losses as high as 70%. Although this is regarded as a relatively high figure, bait loss caused by birds should be recognised as a factor that can significantly reduce the longline catching efficiency. Different approaches for solving the problem have been tried, from scaring the birds away with visual and acoustic stimuli, to luring them from the line by discarding fish liver and other offal when setting the gear. Putting more weights on the line to obtain a faster sinking speed has also been a common way of reducing this type of bait loss.

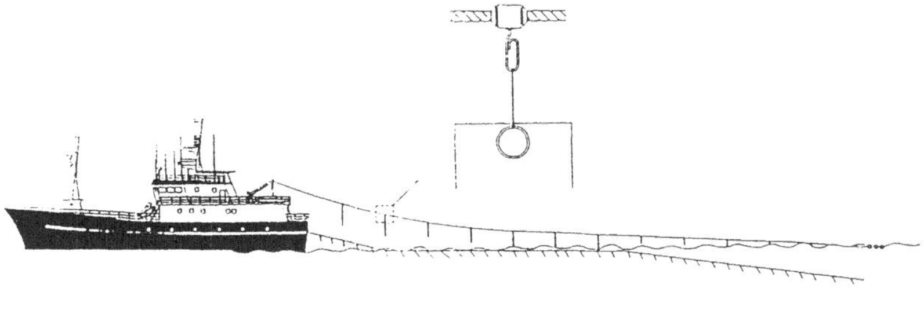

Fig. 5.11 Bird feeding zone during setting of longline and bird scarer. (After Løkkeborg & Bjordal, 1992b.)

One of the more effective methods has been the development of a bird scarer made from a line with pendants at certain intervals that is towed over the longline during setting (Fig. 5.11). This device, which was originally developed to scare albatrosses away from the longline, has also been proven to give a significant reduction in bait loss caused by birds when tested in the Norwegian longline fishery.

The final solution to this problem is to guide the longline beneath the sea surface directly behind the stern of the vessel so that the baited hooks are made physically unavailable to the birds. This has been practised by some Japanese longline vessels, where the line is set through a tunnel in the hull constructed such that the line pays out under water. Another recent development that has been tested with good results is an outboard funnel that guides the line from the setting position on the stern down to 1–2 m depth closely behind the vessel (Fig. 5.12). The setting funnel is hinged to the stern of the vessel and during steaming or hauling it is 'parked' against the stern. During setting it is swung (by means of hydraulics) into the setting position. To allow larger items to be set when using the funnel, it has a longitudinal slot so that, for instance, anchors can be dropped beside the funnel, with the anchor line sliding down through the slot.

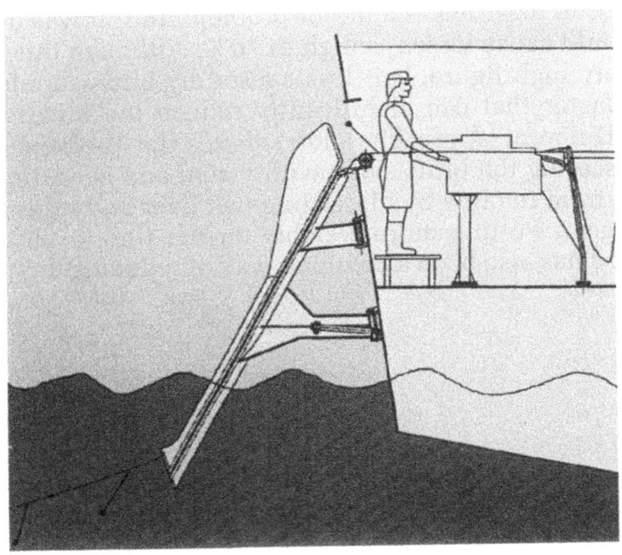

Fig. 5.12 Longline setting funnel. (After product brochure of Solstrand Shipyard, Norway.)

After the line has sunk out of the range of birds there is believed to be little bait loss until the line has reached its fishing depth. For pelagic and semipelagic longlines, bait loss is thought to be not much more than that caused by the target and potential bycatch species. For bottom set lines, however, further bait loss may be caused by different benthic animals, here referred to as bait scavengers. As their abundance and diversity vary with depth and type of seabed, different bait scavengers may cause bait loss varying from slight to severe. Typical scavengers are hagfish, sea lice (amphipods, isopods), starfish, sea cucumbers, sea urchins, crabs and other decapod crustaceans: animals that most often steal the bait without normally being hooked themselves. At some fishing grounds, longline fishing may be severely hampered through bait loss to scavengers as most of the bait may be eaten before any significant number of target fish have been attracted to the line. There are no good solutions to this problem. However, in some fisheries, small floats with 3–10 g buoyancy have been used on each snood in order to lift the baited hooks off the bottom and thus make the bait unavailable to most bait scavengers. The use of small floats to lift the baited hook was common in the earlier Swedish longline fishery for ling in the North Sea, with a rigging of the floats on the snoods so that the baited hook was lifted about 1 m above the sea bed. This type of rigging, which requires long snoods and correspondingly wide hook spacing, is not practical with modern (mechanised) longlining. Trials were therefore conducted with a type of float (Fig. 5.13) that was moulded on to standard snoods (0.5 m), and significant catch improvements were obtained for ling. However, when evaluated against the additional cost of the floats, the catch improvement did not justify their use in commercial longlining.

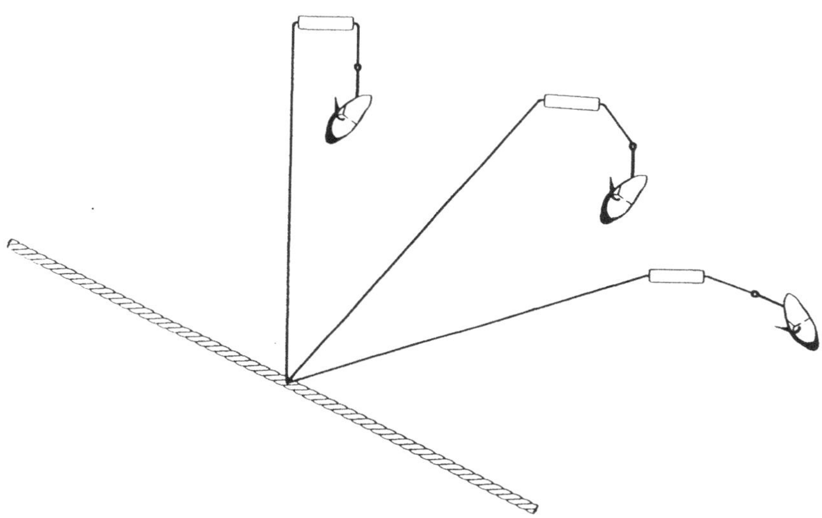

Fig. 5.13 Floats moulded on to snoods: from flume tank tests to investigate the lifting power of the floats with increasing current.

A phenomenon often experienced by fishermen is that of 'feeding the scavengers'. If the lines are set several times on the same ground only poor catches and a high bait loss are obtained initially. However, after some time the catch rates improve and bait loss is reduced. The usual explanation for this is that it takes a few sets to satiate the scavengers and then the bait is made available for the fish.

Hooks that have survived the setting and first period at fishing depth with the bait still intact represent the actual effective hooks available for catching fish. However, the number of effective hooks for the target species and sizes of fish sought may still be reduced through inter- and intraspecies competition for the baited hooks. As the prime target catch is large individuals of one or a few species, every hook being occupied by a non-target species or undersized target fish will reduce the catch rate and profitability. This cannot be avoided, but the negative impact by competition from non-target species can be reduced. Careful positioning of the gear by the skilful and experienced skipper (also considering displacement of the lines by currents when sinking) is important to avoid unwanted catch. Further, there are clear possibilities for improving the species and size selectivity of the gear through the choice of bait type as well as hook and bait sizes (see Chapter 6).

When a target fish has been attracted to and has ingested the baited hook, the role of the bait has been fulfilled. Now successful capture depends on the hook and snood, first, on the effectiveness of the hook to get the fish hooked, and secondly on its ability to retain the fish throughout the soak time and during the retrieval of the gear. It is believed that there is little loss of hooked fish at fishing depth. However, when the lines are hauled from depths down to more than 2000 m, there is a clear risk of losing fish. Fish loss during hauling of longlines has never been quantified, mainly because of the lack of a proper observation methodology. However, with ideal weather and sea conditions it is possible to observe fish being hauled through the water column by the echosounder. Figure 5.14 shows a typical recording, where the weak traces represent individual hooks without any catch. The more marked traces are echoes from hooked fish being hauled through the water column. In this example it can be seen that one fish

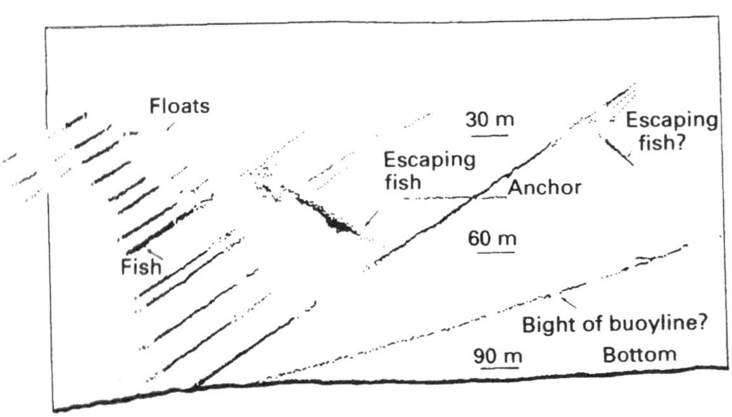

Fig. 5.14 Examples of echoregistrations of different targets during hauling of longline.

escaped and swam downwards after being lifted off the sea bed. Hauling longlines in bad weather conditions and strong currents increases the risk of losing fish, as the rolling and pitching movements of the vessel are transplanted down the line. Naturally the risk of fish loss increases with increasing fish size as this applies stronger forces between the fish and the hook. It is not possible to avoid fish loss during hauling, but the problem can be minimised in two ways. First, it is important to choose hooks that hold the fish well (e.g. Circle, EZ-baiter) and to ensure careful maintenance of the gear, particularly by regular replacement of snoods, even if they are only slightly damaged. Secondly, it will normally be clearly advantageous to slow down the hauling speed somewhat during unfavourable weather and current conditions, as experience has shown that this does reduce the fish loss.

Chapter 6
Selective Properties of Longline Gear

6.1 *The importance of gear selectivity*

The act of fishing and the selectivity of longlines involves four stages (Fig. 6.1). First, the area fished and the ranges of the fish must overlap in space and time, i.e. the fish must be available within the fished region. Secondly, the fish must encounter the operating range of longlines, i.e. the fish must be accessible within the odour plume (active space) generated by the baits. Third, the fish must respond to the bait odour

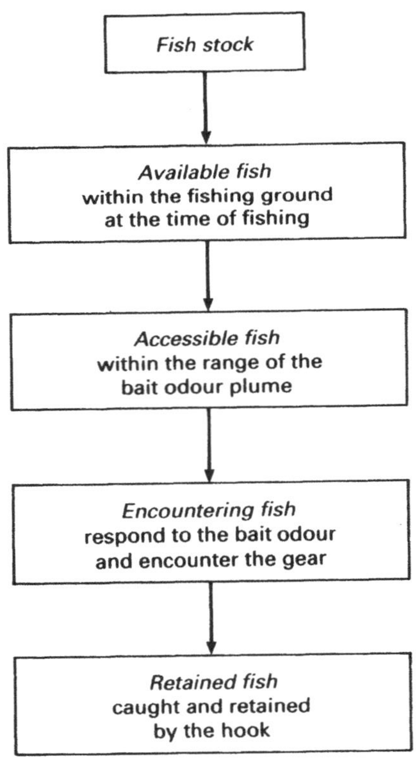

Fig. 6.1 The four stages of the longline catching process.

and locate the baited hooks, i.e. the fish must encounter the gear. Finally, the fish must attack the baited hooks, get hooked and be retained by the hook.

Individual fish are not equally vulnerable to any type of fishing gear because they are heterogeneous in species, size, behaviour, habitat, etc. Hence the species and size compositions of fish vary between catches taken with different gear types and differ from the actual species and size composition on the fishing ground. *Selectivity* is the term given to the process that causes the probability of capture to vary with the characteristics of the fish; longline selectivity may be defined as the proportion of each species and size of fish caught and retained from those present on the fishing ground.

Selective fishing is strictly speaking a paradox because its reciprocal concept, non-selective fishing, does not exist. All forms of fishing are more or less selective because the catching process of the gear itself and the way it is operated cause some species and sizes of fish to be caught more effectively than others. For example, longlines catch haddock more efficiently than cod because haddock more often attack a stationary prey such as a baited hook (see Figs 2.8 and 2.9) and a gillnet of a given mesh size is most efficient in catching fish of a size corresponding to the mesh opening.

In commercial fishing the fishermen often want highly selective gear to target certain species and reduce to a minimum the bycatch of unwanted species and sizes. The groundfish fisheries in the Barents Sea are managed by bycatch limits for undersized fish and species that are fully exploited (e.g. cod) or overexploited (e.g. Greenland halibut). When targeting other species in these mixed fisheries the fisherman therefore needs a gear that is both size and species selective. The management measures for this region ban vessels from operating on a fishing ground if the legal proportions of undersized fish or overexploited species are exceeded. In the north Pacific Ocean, the groundfish fisheries are managed with strict bycatch limits for halibut and king crab, for example, which may result in the closure of some fisheries before the total allowable catch (TAC) of other target species is taken.

Thus, knowledge of fishing gear selectivity is important for the proper management of fish stocks because very few fisheries are based solely on a single species. The fisherman will benefit from an understanding of how the gear configuration affects the size and species of fish harvested, and can thereby design gear that catches the target species more efficiently and protects young fish and overexploited species.

6.2 Species selection

Fishing gear selectivity is influenced by four main factors that reflect the stages of the fish capture process described above. Different species and sizes of fish show preferences for different areas and depths, i.e. habitat preferences, and their availability and accessibility to any given fishing operation will differ. The outcome of a fishing operation will therefore be influenced by the area and depth at which the skipper chooses to set his gear. Once the gear has been set at a certain fishing ground and water depth, the resulting catch composition is affected by the performance of the gear and the response of the fish.

Different species and size groups of fish are to a certain degree separated in space and time, and a knowledge of species and size-dependent habitat preferences is an important basis for selective fishing. From experience the skipper has learned that the species composition of the catch is affected by the species of fish available within the region fished at the time of fishing. In fishing for groundfish, for example, the catch composition attained for lines soaked at the continental shelf is different from lines soaked at the continental slope and this also changes seasonally. When operating at a fishing ground the accessibility of different species to the gear is determined by the bottom substrates and the water depth at which the lines are soaked. As exemplified by the tuna fishery, the compositions of pelagic species in longline catches are greatly affected by the fishing depth. Differences in the vertical distribution of tuna may be a reflection of preferences for different water temperatures. There are also diel (daily) patterns in the depth preferences of tuna and other fish. A better understanding of fish behaviour would therefore be beneficial to fishermen since it would enable them to choose the best regions, substrates, depths and times at which to set their longlines.

As different species usually inhabit the same area at the same time, few fisheries are based on one species only. To be able to fish selectively for the target species in a mixed species fishery, the gear has to be designed to catch the target species more effectively than the unwanted species. Knowledge of fish behaviour is of great importance in developing species-selective fishing gears because the outcome of a fishing operation is the result of a complex interaction between gear performance and fish behaviour. Fishing with longlines takes advantage of a fish's demand for food, and the catch result is therefore determined by interactions between gear parameters, particularly bait parameters, and feeding behaviour.

Bait type is regarded as the most important parameter affecting species selectivity in longlining. This is explained by different species of fish having different preferences for food items. Analyses of the compositions of prey in the stomach of fish of different species caught in the same area have shown that cod and haddock, for example, feed on different types of prey. Further, behaviour studies in the laboratory have shown that fish are able to select different types of food based on their different smell and taste. This behavioural aspect of feeding is the basis for using different bait types for catching different target fishes, and explains why, for example, mackerel- and squid-baited longlines give catches that have different species compositions.

There are numerous examples, both from scientific investigations and from fishermen's experiences, that show that the catching efficiency of a bait type is different for different species. In the mixed longline fishery for tusk and ling in the north eastern Atlantic, it has been shown that squid bait catches twice as many ling as mackerel bait, but only 9% more tusk. Mackerel bait appears to be most effective for catching haddock, whereas squid is found to be most effective for the capture of cod. In the tuna fishery, saury has been shown to be more effective for capturing tuna, and mackerel more effective for marlin.

Work aimed at developing artificial baits has also demonstrated the species-selective effects of baits. A longline bait based on minced fish products reinforced by a netting gave catch rates of haddock that were several times higher than that of natural mackerel bait, whereas the catch rate of cod for this bait type was slightly lower than that of mackerel bait. This result is very interesting for the longline fishery for cod and haddock in the Barents Sea, as the cod stock in this area is at present more strictly regulated than the haddock stock. It is therefore desirable to fish more selectively for haddock and relatively less from a restricted individual cod quota. Regarding the species-selective effects of bait type it is important to note that fish may change their bait preference seasonally and that previous diet experience also may affect their food preference. Thus, the bait which has proven to be most effective in catching the target fish in one fishery may not be the most effective bait in other seasons or areas.

The type of hook also affects the species composition of longline catches. This is due to the fact that fish caught by baited hooks are hooked either in the mouth cavity, particularly in the jaw, or in the alimentary tract if the hook is swallowed. Some species are hooked more often in the mouth than others. This is mainly the case with more discriminate feeders, which do not swallow the bait directly after ingestion but tend to keep the bait in the mouth to taste it and test its structure. When doing this, the fish often try to move away and are thus often hooked in the mouth as the snood is stretched. Behaviour studies and longline fishing experiments have shown how hooks can be designed to penetrate the tissue of the mouth cavity more efficiently than the traditional J-shaped hook (Fig. 6.2). These hooks therefore more efficiently catch species that are typically hooked in the mouth and may be used to fish selectively for such species. Halibut, haddock, ling and hake are mainly hooked in the jaw, and the Circle hook which is used in the longline fishery for halibut, has in comparison with the J-hook proven to be superior in catching halibut and hake. Cod and tusk which also swallow the baited hook, were caught in equal numbers by the two hook types. In a similar comparison, the EZ-baiter circle hook, which has an intermediate shape between the Circle hook and the J-hook, gave 50%

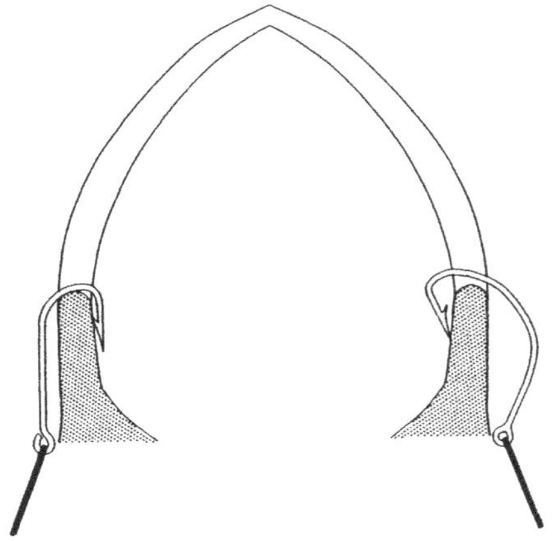

Fig. 6.2 The angle between the fish jaw and the hook point of the J-hook and the Wide Gap hook, showing that the latter is more likely to penetrate the tissue of the mouth cavity. (After Huse & Fermö, 1990.)

higher catch rates for haddock (mainly hooked in the mouth) than the J-hook, but only 20% higher catch rates for cod. These fishing experiments demonstrate that the hook may be designed in accordance with the hooking behaviour to selectively catch certain species.

6.3 Size selection

Fishing strategy with respect to the choice of fishing ground and water depth affects not only the species selectivity but also the size selectivity of longlining because different size groups of a species may show different horizontal and vertical distributions. The distribution patterns of fish of different sizes are used by fishermen to target larger fish. Fish distribution, however, shows diel cycles and seasonal changes, and a deeper understanding of these mechanisms is needed for improved fishing strategies to avoid catching small fish.

Baited longlines catch fish that are searching for food, and, as mentioned earlier, baited hooks may attract fish from an extended distance of several hundred metres. Larger fish swim faster and are therefore searching for food over a larger area than small fish. Hence there is reason to believe that a higher proportion of large than small fish will locate the longline. The difference between small and large fish in their ability to locate baited hooks is likely to be most pronounced for fish at a longer distance from the longline, i.e. for fish occupying the distant part of the area influenced by the odour plume.

Once the fish have located the longline there is competition for the available baits. Behaviour studies using underwater video cameras have shown that large fish frighten and chase away smaller fish from the baited hooks. These observations indicate that the largest individuals are the most successful competitors among fish that are attracted to baited longlines. Thus, differences in swimming speed and competitive capability between small and large fish may cause a selection process that brings a relatively high proportion of large individuals into contact with the gear. This process, which is not influenced by the gear performance itself, is typical for baited gears, and does not affect selectivity in active gears such as trawls.

The main gear parameter that influences longline size selectivity is the bait size. Fish feeding in a habitat with a mixture of natural prey of different sizes will often show a preference for certain prey sizes. Smaller fish generally prefer prey below a certain size, as determined by such factors as their mouth size and their ability to capture and handle the prey. Therefore, bait size is likely to be an important factor affecting the size of fish caught by baited hooks. This effect has been demonstrated in fishing experiments using baits of various sizes. Large baits caught fewer cod below 60 cm length than small baits, whereas large and small baits caught the same number of fish larger than 60 cm (Fig. 6.3). The same tendency has been demonstrated for haddock, but the size-selective effect was less pronounced for this species as small baits also gave higher catch rates for larger fish (see Fig. 5.4). This interesting difference between cod and haddock can be explained by their different behavioural responses towards baited hooks. Cod most often suck the whole bait into the mouth, and small cod will not respond to and attack large baits. Haddock, on the other hand, typically nibble small pieces of the bait while the rest of the bait and the hook are outside the mouth. In this way smaller haddock also attack large baits and try to bite small parts of it without becoming hooked. Because haddock makes several

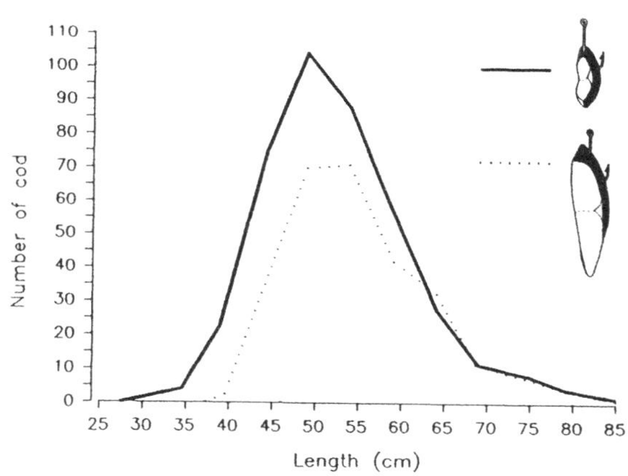

Fig. 6.3 Comparison of the length–frequency distributions of cod caught on mackerel baits of two different sizes, 10 g (solid line) and 30 g (dotted line), respectively. (Modified after Johannessen, 1983.)

attacks on the same bait, the bait becomes smaller and smaller and the fish may finally bite the hook and be caught.

Reducing the proportion of small fish in longline catches by increasing the bait size does, however, also increase the total bait consumption and thereby the bait cost. An interesting solution to this problem has been sought by using an inedible body on longline hooks (Fig. 6.4). A plastic body attached to the hook shank may, in combination with a normal or even small bait, appear as a large item to the fish. This combination has been shown to give the same size-selective effect as large baits in pelagic longlining for haddock. Hooks with a plastic body and mackerel bait caught a lower proportion of undersized haddock (< 44 cm) than hooks with mackerel bait only. The same effect was not obtained in experiments with bottom longlines for tusk and ling. This is presumably because the fish do not get a good visual impression of the baited hook lying on the bottom, particularly in deep water with very low light levels.

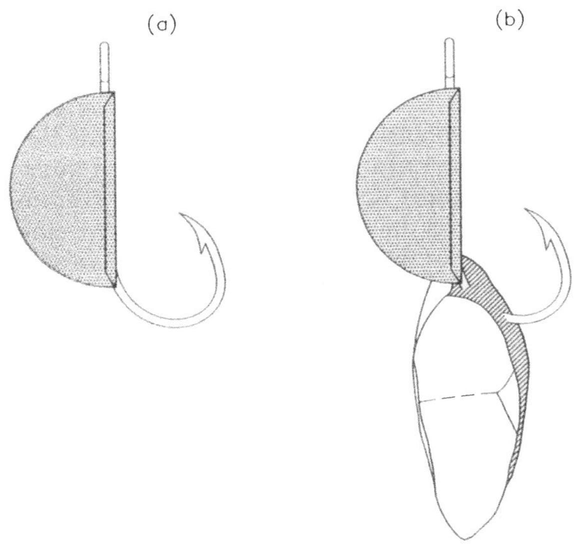

Fig. 6.4 (a) Hook with a plastic body attached to the shank. (b) The plastic body in combination with a mackerel bait of normal size appears as a large item. (Redrawn after Løkkeborg & Bjordal, 1995.)

The hook size is also believed to influence the size of fish caught by longlines, i.e. larger hooks catching larger fish. However, there is no clear evidence for a size-selective effect of hook size. The explanation for a potential effect of the hook size is that larger hooks demand a stronger force to fully penetrate the tissue in the mouth cavity, and larger fish may exert a stronger tension on the snood when they make jerks and rushes with the baited hook in the mouth. On the other hand, the mouth cavity of smaller fish has thinner skin that is more easily penetrated by the hook point, and small fish may therefore be hooked with less tension on the snood than large fish. Another potential explanation for larger hooks catching larger fish is the mechanical strength of the hook. Larger hooks are less readily broken or straightened and may therefore be more effective in preventing large fish from escaping during retrieval of the line. Today the hook manufacturers use alloys of high quality to make hooks with high breaking strength and elasticity, and small hooks can therefore hold fairly large fish. Thus, it is reasonable to conclude that at least moderate differences in hook size (i.e. within the range that is relevant to commercial longline fishing) have little effect on the size composition of longline catches.

Chapter 7
Longlining Compared with other Fishing Methods

Why are so many capture techniques applied in fishing, even for the same species in the same area – and why do some fishermen stay with one fishing method while other fishermen in the same fishing community stick to another gear type? These are questions that need a thorough sociological study to be answered – if the reasons can be fully clarified at all. However, some key words for the explanation of the highly varied fishing practices are: the historical development of gears; the heritage of traditional cultures and practices; the inherent conservatism in fishing communities; capital available for investment; and, in more recent years also, fisheries regulations, fisheries development schemes and public opinion effects. Thus there are various reasons why longline gear is predominant in some fisheries and little used in others. In the following we will discuss some of the factors that favour the use of longlines in comparison with other gears, and consider factors of importance for future trends.

7.1 The relative catching efficiency of longlining

When compared with most other fishing gears (i.e. trawls, seine nets, purse seines, gillnets, jigging) longlining appears to be a relatively inefficient fishing method. This is clearly true with high fish densities, as the longline is one of the few fishing gears of the world that has a definite saturation point – limited by the number of hooks that can be set and hauled during a day. The daily catch of a longline vessel therefore has an upper limit that is defined by a maximum number of fish, being equal to the daily number of hooks hauled multiplied by the average weight of the fish. However, the catch per unit of effort (CPUE) is normally far below this maximum level. The daily catch potential (C_D) of a longline operation can be expressed by the following equation:

$$C_D = N \times B_E \times B_B \times B_S \times B_N \times C_T \times W_T,$$

where

N = the number of hooks set and hauled during 24 hours

B_E = the baiting efficiency (fraction of hooks leaving the vessel with bait on)

B_B = bait loss caused by sea birds during setting (fraction of baited hooks remaining after bait predation by birds)

B_S = bait loss due to scavengers (fraction of baited hooks after bait predation by bottom scavengers)

B_F = bait loss caused by target and non-target fish, including non-target fish that are hooked, and thus preventing capture of target fish*

C_T = the availability and catching probability of target fish (related to the number of baited hooks available for target fish = number of *effective hooks*)

W_T = the average weight of target fish

Applying this catch equation to a practical example of an autoline vessel fishing for cod, it can be shown that the real saturation level of the longline gear is normally far below the actual number of hooks operated during a day. If 30 000 hooks are set, the theoretical catch of large target fish (e.g. cod with average weight of 5 kg) would be 150 t. However, this is of course a purely hypothetical figure as the catching efficiency is reduced through several steps according to the above equation. The baiting efficiency is normally about 85%. Thus 25 500 effective hooks are set from the vessel. The number of effective hooks is further decreased due to bait loss from sea birds, bait scavengers and non-target fish. If the bait loss from these sources is set to 15%, 15% and 5%, respectively, the number of effective hooks available for the target fish is reduced to 17 500. Using a catching probability for target fish of 0.2 and an average fish size of 2 kg (de-headed and gutted), this gives a total daily catch of 3500 fish with a total catch weight of 7000 kg. This corresponds to a fairly good average daily catch rate of cod for an autoline vessel, although it may vary from a few tonnes up to 15 t. Thus, the catching power of a longline operation increases with the number of effective hooks and the availability (density) and average size of the target fish.

With medium fish abundance, a daily catch of 7 t could also be a normal catch rate for a trawler or a gillnet vessel. However, with increasing fish densities, the longline will meet its saturation level, where the absolute maximum number of fish caught equals the (limited) number of effective hooks. With trawl or gillnet fishing, the catch will increase proportionally to the increase in fish density. So from normal daily catch rates of 10–20 t for a cod trawler, this can be obtained during one hour of fishing at very high fish densities. With such high catch rates the limit to the daily catch is not related to the gear but is more restricted by limitations in the capacity to handle the catch. Particularly with active fishing gears, these can be set again and again at high fish densities and thus produce very high daily catches. With

* The term 'target fish' does include fish of marketable species and size.

increasing fish densities, longlining is therefore clearly a relatively inefficient fishing gear.

On the contrary, with scattered, low density fish distributions, longlining has a high catching efficiency when compared with most other fishing methods. If the example of an autoline vessel fishing 30 000 hooks daily is used, fish are attracted from a large area as compared with the swept area of a trawler. With a hook spacing of 1.4 m, and given that the longline attracts fish from a distance of 800 m (see Section 2.3), the fished area is nearly 34 km^2. In comparison, the area swept daily by a trawler (based on 18 hours dragging the trawl at 3 knots and 50 m sweeping width) would be about 5 km^2. The longline's ability to attract and aggregate fish from a large area therefore makes it a relatively efficient gear with moderate to low fish densities.

7.2 Responsible fishing

The recognition that global marine resources are limited, that the present catching capacity of the world's fishing fleets far exceeds what is needed to harvest the surplus production, and that different fishing methods have different undesired side effects on the resources and the environment has recently put the focus on responsible fishing. Responsibility in this context means that measures should be taken to safeguard a sustainable harvest from aquatic resources so that the negative impact on the environment is minimised. In a regime for responsible fishing there are three main elements:

- Reliable scientific advice on status and development (growth, decline) of fish stocks
- Exploitation rates (quotas, TACs) in accordance with scientific recommendations
- Fisheries regulations and management control measures to ensure that quotas are not overfished and that the fishing activities do not have significant negative effects on the biological resources or their environment.

The first and last elements require considerable effort and adequate methods and equipment, including extensive at-sea monitoring of the fish stocks and the fishing activities. The TAC element requires responsible attitudes and political courage within and between nations to keep fishing quotas within recommended exploitation levels. However, the control task can be simplified by the use of more responsible fishing techniques, as more selective and gentle fishing methods do, for example, reduce the need for extensive monitoring and control of the bycatch and discards.

We should now consider how longline fishing compares with alternative gears regarding responsible fish capture. An extensive assessment of this issue is beyond the scope of this book. However, to illustrate some basic differences with respect to responsibility between alternative gears a comparison is made between longlining, gillnetting and trawling: evaluating such factors as selectivity, unaccountable fishing mortality, ghost fishing, effects on the sea bed, relative energy consumption and pollution.

Size selection

Among the three gears the gillnet is clearly the most size selective. Gillnet catches normally have a narrow size distribution. Catching the desired size groups of fish with gillnets is brought about by choosing the corresponding mesh size. Most of the small fish avoid being caught as they swim through the meshes, and if not entangled, the larger fish do not get caught as they are too big to enter the mesh.

In trawling, size selection takes place at different stages of the capture process, from the doors and sweeps to the codend. However, for small fish, codend selection (or really escapement through the codend meshes) is most important. In principle, mesh selection in trawls should produce as good as or even better size selection of fish than can be obtained by gillnets as trawls also retain the largest fish that are not caught successfully by gillnets. In contrast to gillnets, where the meshes are kept open, trawl codend mesh selection may be hampered in different ways. The meshes may be clogged by fish (particularly spiny fishes or flatfish) or may be stretched (closed) for a variety of reasons so that the size-selective properties are reduced.

In contrast to trawls that capture most of the fish in front of the trawl opening so that selection must take place after the fish has entered the gear, much of the size selection in longlining is based on a two-step behaviour-dependent process: first when the fish is searching for the gear, then during competition for the baited hooks. Both these processes tend to favour or select for larger fish. Thus, the general capture principle of longline fishing is size selective when fishing on grounds with a normal (wide) size distribution of fish, as the larger fish do 'outcompete' the smaller fish. However, it should be noted that longlines can technically catch large quantities of small fish if there are mainly small fish on the fishing ground. In practical fishing, though, this is unlikely to be the case, owing to the limited catching power of longlines with an upper limit regarding the number of fish that can be caught per unit of time. As fishermen are paid by the weight of marketable fish and generally with significant higher prices for the larger fish, every small fish on a longline has a direct negative effect on profitability. Thus, the catch weight decreases linearly with an increasing proportion of small fish in longline catches, and the catch value decreases with an even higher factor. With too high proportions of small fish in the catches, the longline skipper therefore has the choice to either stop fishing or search for grounds with higher proportions of large fish. The direct dependency between fish size and profitability in longlining is therefore a strong incentive towards exploitation of larger fish. The profitability of trawl fishing is also reduced with an increasing number of small fish in the catch. Compared with longlining there is, however, an important difference based on the differences in catching power between the two gears, where increasing proportions of small fish in trawling do not directly reduce the number of large fish in the catches. As long as trawl catches contain adequate amounts of large fish to make the fishing profitable, there is no incentive to stop fishing or change fishing ground, even if the catches contain large quantities of smaller fish, which are either marketed at low prices or discarded. In contrast to longlining there is therefore no strong inherent mechanism in the operational tactics of trawling against the exploitation of small fish.

Little information is available on the actual size composition of catches taken by trawl, gillnet and longline targeting the same species on the same fishing ground.To illustrate the relative differences in size selection between the three gears, Fig. 7.1 shows the size distributions of Greenland halibut taken by trawl, longline and gillnet during simultaneous fishing in the same area. Gillnets were clearly most selective for larger fish: the trawl had poor size selection while the longline catches had an intermediate size distribution. Besides illustrating the comparative selection properties of the gears, this example also underlines the differences in catching power and operational tactics with respect to profitability. As shown in the figure, all gears caught about the same quantities of large fish (i.e. above 60 cm). However, the trawler also caught considerable quantities of smaller fish as compared with the longline and gillnet vessels.

A similar difference in catch size distribution between these gears has also been a pronounced long-term trend in the cod fisheries off northern Norway. Over a 10-year period, the proportion of small cod (below 45 cm) in trawl landings was close to 20%, while the proportions of small fish in the longline and gillnet catches were about 6% and 1%, respectively (Table 7.1).

Another example of a similar exploitation pattern can be found in the multigear fishery for hake in the Gulf of Lions (Mediterranean), which is conducted by French trawlers and gillnetters and by Spanish trawlers and longliners. With codend mesh sizes of 38–40 mm and gillnet meshes of 80 mm, the resulting size distributions in the catches are as given in Table 7.2. Similar differences in size selection between longline and trawls have also been demonstrated in the hake fisheries off southern Africa (see Fig. 4.1).

These stocks are characterised as overfished, and a change towards exploiting larger fish groups could have been obtained by increased mesh size in the trawls or a shift in fishing effort from trawling to gill-

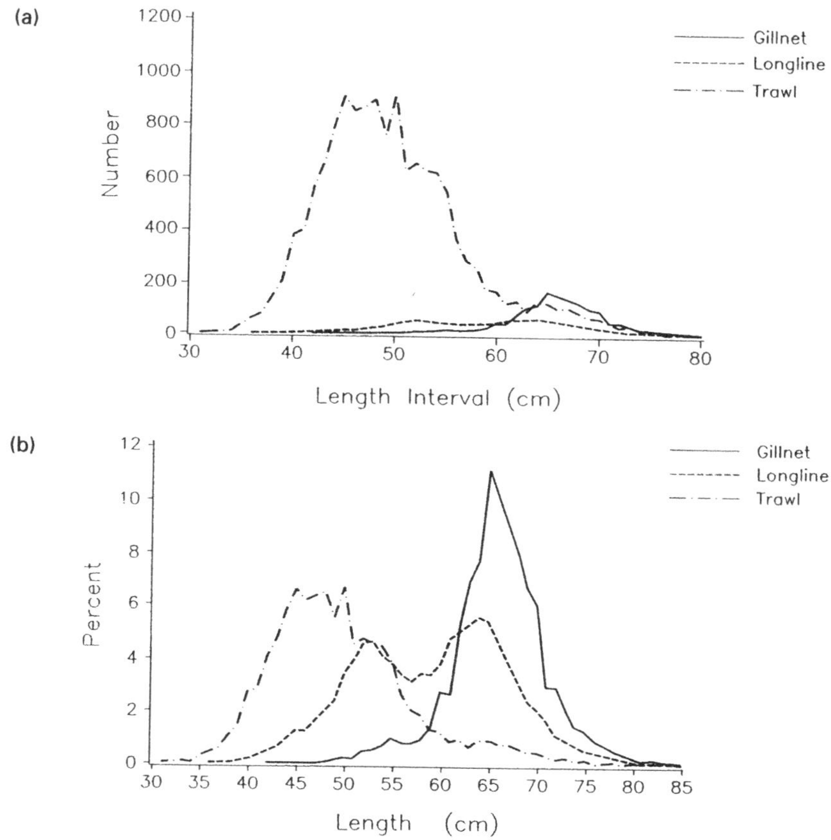

Fig. 7.1 (a) Absolute and (b) relative size distributions of Greenland halibut from comparative fishing with trawl, longline and gillnet. (Adapted from Nedreaas *et al.*, 1993.)

Table 7.1 Landings of cod by different fishing gears in northern Norway, during the 10-year period 1978–1987, including the proportion of smaller fish (< 45 cm length, de-headed). Adapted from Bjordal (1989).

Gear	Total catch (tonnes)	<45 cm (tonnes)	<45 cm (%)
Longline	178 667	10 853	6.1
Gillnet	350 070	4 659	1.3
Trawl	399 419	77 304	19.4

Table 7.2 Mean lengths (cm) of male and female hake in a Mediterranean multigear fishery (min. and max. lengths are given in parentheses). Adapted from Aldebert *et al.* (1993).

Gear	Males	Females
Longline	44.7 (29–60)	54.3 (29–85)
Gillnet	41.7 (27–67)	44.2 (27–86)
French trawl	22.6 (10–66)	22.6 (10–88)
Spanish trawl	19.2 (7–60)	16.7 (7–65)

netting or longlining. This is not easily done as its effects might have severe short-term social implications. However, for long-term responsible management of this stock, such measures should be imposed gradually in this multigear fishery.

Comparing the selective properties of trawl, trap, longline and set-net gears in the fishery for sablefish off the US west coast, Klein (1986) found similar trends in selectivity:

> Although trawls are very effective for harvesting groundfish, particularly rockfish and flatfish, trawl gear appears to be the least desirable gear for harvesting sablefish – trawlers capture the smallest-sized fish, receive the lowest prices and trawls are the least selective gear for sablefish. The remaining three gear types have distinct advantages and disadvantages. Traps are the most selective gear for sablefish and have the highest landing rates, but capture more smaller fish than longlines or set-nets. Set-nets are an effective gear, harvesting the largest-sized sablefish, but landings have been less than 50% sablefish in the last three years. Longlines are fairly selective for sablefish, receive the highest prices per pound and capture primarily large fish.

The examples given above are in general representative of the difference in size selection between longlines, gillnets and trawls, although the magnitude of the difference may vary between fisheries and the relative difference may be altered by management measures such as mesh size regulations. However, the recent development of sorting grids in trawls has also improved the size selectivity in some trawl fisheries, and might, through further development and implementation in different fisheries, contribute to a less pronounced difference in selection properties between trawls and traditionally more size-selective gears.

Species selection and discards

None of the three gears can be classified as particularly species selective. Considering that trawls catch most of what appears in front of the trawl mouth, the basis for species selectivity is generally poor as the gear catches everything that is retained by the codend meshes regardless of species. Likewise, longlines can in principle catch any species that is attracted to the baited hooks, and gillnets any species that encounters the net and matches the size of the meshes so that it is gilled or caught by entangling. However, as in the case of size selection, the limited catching efficiency of longlines means that every fish of a non-marketable species caught reduces the profitability, as it occupies a hook that otherwise could have caught a target fish. The fishing strategy of a longline skipper is therefore more strongly influenced by selecting fishing grounds with few non-target species, as compared with trawl and gillnet skippers since the profitability of the latter gears is not directly hampered by non-target bycatches.

Bycatch of non-marketable species is normally discarded at sea. In the world fisheries, the most severe discard problems are found in the shrimp and prawn trawl fisheries where discard rates of 10–15 kg per kilogram of landed target species are not uncommon. Little information exists on the discard rates of gillnet, longline and trawl fisheries for the same target species. However, the discard per target catch weights in Table 7.3 give an indication of the discard levels in trawl and longline fishing. Even if these discard rates cannot be directly compared, as they are mostly obtained in different fisheries, the top weight-based discard rates recorded for bottom trawling are two to five times as high as the top rates for longlining, indicating that in general discard rates in trawling are significantly higher than in longlining. In the source report, gillnets were not identified among the gears having high discard weights.

In addition to 'normal' fish bycatches, fishing gears also incidentally catch other animals such as marine mammals, sea birds and turtles. This has been strongly criticised by environmental agencies, and this

Table 7.3 The top weight-based discard to landed target catch ratios by gear type. Adapted from Alverson *et al.* (1994).

Fishery	kg discard/kg landed
Demersal trawl	
NW Atlantic fish trawl	5.28
Bering Sea rock sole	2.61
British Columbia cod	2.21
Gulf of Alaska flatfish	2.08
NE Atlantic dab	2.01
Longline	
Bering Sea Greenland turbot	1.03
Eastern Central Pacific swordfish	1.00
Bering Sea sablefish	0.50
Gulf of Alaska cod	0.26

type of bycatch must also be considered seriously when assessing the responsibility of a fishing method. In trawl fishing this is not regarded as a big problem, although incidental catches of sea lions are known, for instance in the Alaska and Namibian trawl fisheries. Gillnets are responsible for bycatches of sea birds, marine mammals and turtles, and there are few measures that can be taken to avoid this. In longline fishing the main problem of this kind is the incidental catches of sea birds such as fulmars, kittiwakes, gannets, great skuas and albatrosses which get hooked when feeding on baits during setting of the lines. Although few birds are actually hooked compared with the number of attacks on baited hooks, a significant number of sea birds may be killed in different longline fisheries. However, with the use of recent developments such as more effective bird scarers (Fig. 5.11), the problem of sea-bird bycatches is strongly reduced and can be eliminated with the use of a setting funnel (Fig. 5.12).

Survival after escape

The survival of fish after escaping from fishing gears is an aspect of responsible fishing that has been focused upon in recent years. With trawls and gillnets this is mainly related to escape of fish through net meshes and the probable mortality caused by injuries to the skin of the fish from being squeezed through a mesh. It has been shown that pelagic species such as herring may suffer high mortality after escaping from trawls or nets. On the other hand, investigations have shown that codfish such as cod and haddock have high survival rates (80–95%) after escaping through trawl meshes. In a comparison between trawls, gillnets and longlines, the high survival rates of cod-like species is more relevant as these are species commonly fished by each of these three gears.

With gillnets and longlines little is known about fish survival after escaping from the gear. For cod-like fishes there is reason to believe that skin damage caused by gillnet meshes does not induce high mortality based on the results of investigations of trawl escapees. Similarly, injuries caused by hooks on fish escaping from longlines should also seldom be lethal, assuming that this type of injury is comparable to mesh-induced wounds. However, this applies to fish with a swimbladder escaping at fishing depth, and fish without a swimbladder. When deep-water fish with a closed (physoclist) swimbladder are being hauled towards the sea surface when retrieving the longline, the swimbladder will gradually expand, the relative expansion being greater the closer the fish is to the surface. At a certain depth the swimbladder is fully expanded and will in most cases rupture as its internal pressure increases further. The swimbladder gas then evacuates into and inflates the body cavity, and the fish becomes positively buoyant and will find it difficult to descend to greater depths. If the fish is lost relatively deep in the water, it therefore has a high probability of descending and surviving after escape from the gear (see Fig. 5.14). However, if the fish escapes after it has become positively buoyant, it will most probably have great difficulty in descending, but rather ascends to the surface with no possibility of survival. These escapees may be picked up by fishermen but are often lost as they surface too far from the vessel. This type of escape or loss of fish has been significantly reduced by the introduction of improved hook designs and swivels, but might still represent a fairly high bymortality in longline fishing. However, the lack of data means that it is difficult to compare the three gears with respect to survival after escape.

Ghost fishing

Ghost fishing, or fishing mortality caused by fishing gear that continues to catch fish after it is lost, is another important issue associated with responsible fish capture. Trawls, longlines and gillnets are incidentally lost due to broken warps and lines when the gear gets stuck on the bottom. Additionally, when fishing at great depths and/or in strong tides, longlines and gillnets get lost because the marker buoys and floats are submerged by the drag from the buoy line. With trawling and longlining, ghost fishing is not regarded as a serious problem. Trawls are seldom lost. However, in cases of gear loss the fish kill is limited to the amount of fish in the trawl and there is no further catch and killing of fish. Likewise, fish kill by lost longlines is limited to the initial number of fish caught. The gear does not continue to fish, as the bait is either eaten by scavengers or it deteriorates within a short time.

In contrast to longlines and trawls, ghost fishing may be a severe problem with gillnet fisheries. As with the other gears, the fish initially caught in lost gillnets will die. However, in contrast to longlines and trawls, gillnets continue to catch and kill fish over a long period after they are lost, because the gear configuration after loss is no different from that of a regular capture situation. Further, fish that die in the net act as bait and thus create a continuous fish aggregation effect around the net. Therefore, ghost fishing catch rates of lost gillnets might be considerably higher than in regular fishing. Lost gillnet recovery programmes have shown that nets can continue to fish for up to ten years after they are lost, and although the total unaccounted mortality due to gillnet ghost fishing has never been quantified, it might be significant in certain fisheries.

Fish quality

Fish quality should also be considered as an aspect of responsible fishing, as fishing methods producing low quality products give sub-optimal resource utilisation. The three fishing methods may all produce first class quality fish products. However, large trawl catches and trawl hauls of long duration might lead to reduced quality as the fish are exposed to high pressure in the codend and with large catches it may take a relatively long time to process all the fish.

Fish caught by longlines are generally considered to be of high quality. The fish come on board one by one, generally still alive, and are bled directly after capture. This gives an excellent basis for high quality fish products. Some line-caught fish may, however, be ruined as a result of attacks from bottom scavengers, a problem that is sometimes experienced in areas with a high abundance of sea lice and hagfish.

Compared with trawling and longlining, fish capture by gillnets has a higher risk of producing low quality catches. Fish caught in gillnets die more easily after capture, and fish quality is therefore often inversely related to the duration of the soak time. Hence, fish caught early in the soaking period may have been dead for hours with correspondingly reduced quality, whereas fish caught later might still be alive when brought on board. In most gillnet fisheries, limited soak time is therefore a prerequisite for high catch quality. However, bad weather conditions and undesirable fishing practices (e.g. the use of a large number of nets, so that individual nets are hauled only every 2–5 days) often lead to long soak times, resulting in a relatively large proportion of poor quality catches.

Energy consumption and pollution

The energy used to catch a unit amount of fish is also an issue that should be considered when evaluating the responsibility of fishing methods. The pollution caused by emission of combustion gases (CO, CO_2, NO_x) is now the focus of concern on a global scale, and the fishing industry is expected to take its share of responsibility in efforts to reach general objectives on the reduction of gas emissions to the atmosphere. However, as emission is an accompanying effect of the relative energy consumption in fish capture, the comparison between the three gears refers to energy consumption only.

The relative energy consumption in fish capture is often expressed as the energy coefficient, which is the amount of fuel oil needed to capture 1 kg of (undressed) fish. Average energy coefficients for demersal trawling, longlining and gillnetting in the Norwegian fisheries are given in Table 7.4.

Table 7.4 Energy coefficients (kg fuel oil per kg (undressed) fish caught), for longlining, gillnetting and trawling, Norway, 1989. (Adapted from Bjørkum, 1993.)

Fishery	Energy coefficient
Longline (coastal)	0.21
Longline (high seas)	0.38
Gillnet (mainly coastal)	0.30
Trawling (wetfish)	0.79
Trawling (factory)	0.81

In multigear fisheries for similar target species, there is reason to believe that this example is representative of the general difference in relative energy consumption between the gears. The energy coefficients of trawling are thus about three times those of longlining and gillnetting. With all three fishing methods, close to maximum engine load is used when steaming to and from fishing grounds. However, during fishing, which constitutes about 80% of a total fishing trip, the trawler's engine still runs at high load compared with a 20–40% load of the relatively smaller engines used in longline and gillnet fishing. The trawl fishing principle thus implies a significantly higher energy cost per unit catch with correspondingly high pollutant gas emissions.

It should also be noted that high-seas fisheries (both longline and trawl) have higher energy consumptions than coastal fisheries. This is mainly because high-seas fishing vessels are larger, with larger main engines and also a significant auxiliary engine capacity to produce energy for the freezing of the catch, for heating the crew's accommodation and to run a wide range of mechanical equipment and instruments.

Among the three gears longlining is the only one using baits. Bait consumption should therefore also be included in the total energy budget. This is not easily included in the energy coefficient, but a simplified conversion could be obtained by subtracting the bait consumption from the catch. With a bait consumption of 0.2 kg per kilogram of catch, this gives energy coefficients of 0.26 and 0.47 for coastal and high-seas longlining, respectively. These are still comparable to the gillnet coefficient and significantly less than the energy coefficients of trawling.

Impact on the sea bed

Beyond the different environmental impacts mentioned above, concern has recently been raised regarding the environmental impacts on the sea-bed habitats from fishing activities. In this context, the main concern is focused on irreversible changes in the sea-bed structure and associated ecosystems. Longline and gillnet fishing obviously have minor effects on the sea bed, whereas trawling has caused clear effects. Particularly with the introduction of more powerful trawling systems (stronger trawl gear and increased engine power), areas that were not fished by trawl earlier because the bottom conditions were too rough are now being fished, resulting in extended levelling and harrowing of the sea bed. Trawls are as a result accused of ruining benthic ecosystems with corresponding negative effects on biodiversity. On the other hand it is claimed that the harrowing effect actually increases fish production by exposing benthic organisms as food for target species. Neither of the claimed effects are properly documented. In future management trends, however, it seems clear that more emphasis will be put on the precautionary principle, meaning that there will be an increasing concern regarding trawling and other fishing methods that may cause irreversible or long-lasting changes to sea-bed habitats.

Summary: responsible fishing

No fishing gears are ideal with respect to the different requirements for responsible fishing, of which some major aspects are discussed in this chapter. However, in future fisheries and marine ecosystem management, there is strong reason to believe that fishing methods with a high degree of responsibility will be favoured over the less responsible ones. In some cases it is easy to evaluate the difference in, for example, size selectivity or relative energy consumption between two or more alternative gears applied in the exploitation of the same fish resource. In other cases it is more difficult, mainly owing to lack of adequate data for a comparative evaluation. In Table 7.5 an evaluation is made of longline, gillnet and trawl fishing with respect to the aspects of responsible fishing discussed above, using a scale from 3 + to 3 – to designate high to

Table 7.5 Responsible fishing aspects: a general comparison between longline, gillnet and trawl gear.

Responsible fishing aspects	Longline	Gillnet	Trawl
Size selection	++	+++	+
Species selection	++	+	–
Survival after escape	+	+	+
Ghost fishing	++	– – –	++
Fish quality	++	–	++
Energy consumption/pollution	++	++	– –
Impact on the sea bed	+++	+++	–
Total	14+	6+	2+

low responsibility. In this exercise longlining comes out as clearly the most responsible fishing method in, for example, a multigear fishery for mixed demersal species. Although this is a subjective and general comparison between alternative gears, such evaluations of responsibility should be applied in future fisheries management strategies. They should then be applied to specific multigear fisheries, using adequate data for evaluation of the different factors and also with a weighted ranking between the factors (i.e. would size selectivity be twice as important as sea-bed effects in the case being evaluated?). In such specific evaluations, there is reason to believe that longlines will generally be favoured in a multigear management strategy, although there will surely be cases where longline fishing does not necessarily come out as the most responsible alternative method.

7.3 Effects on fish stocks and fisheries employment

During this century there has been a development from a situation where most fish stocks were in a good condition, producing higher yields than could be harvested by the capacity of the fishing fleets, to the present situation, characterised by stocks that are mostly overfished and with a surplus capacity of most fishing fleets. Thus, the general picture of the 1990s is that fish resources are limited and too small to sustain the current level of fishing capacity and the number of fishermen. Recognising that many stocks are overfished does imply that they are underutilised and that proper management can provide higher sustained yields from these stocks. Nevertheless, even after a recovery of currently overexploited stocks and a corresponding rise in total yield, it is most likely that this added production will still not satisfy the current overcapacity of the fishing fleets. There is a widely held view that the fleet capacity and correspondingly the number of fishermen have to be reduced to match the harvestable production from the fish stocks. However, fleets and fleet capacity are general concepts that include all fishing vessels and all fishing methods. In a multigear fishery, trawls are in general more effective than longlines, measured as catch per unit of effort (e.g. vessel days or manpower). However, the higher efficiency in trawling tends to be caused by comparatively larger catches of smaller-sized fish. Trawling and longlining therefore have different effects on the stock being harvested. As trawls tend to fish more young and partly undersized fish, trawling implies a much larger risk of growth overfishing than does longlining (i.e. the fish is harvested before the growth potential is fully utilised so that the yield or biomass production from the stock is at a suboptimal level). The possible effects of harvesting a stock alternatively by trawls or longlines are shown in Fig. 7.2. Based on the differences in size selectivity between longlines and trawls in the fishery for Pacific cod in the Bering Sea, a numerical

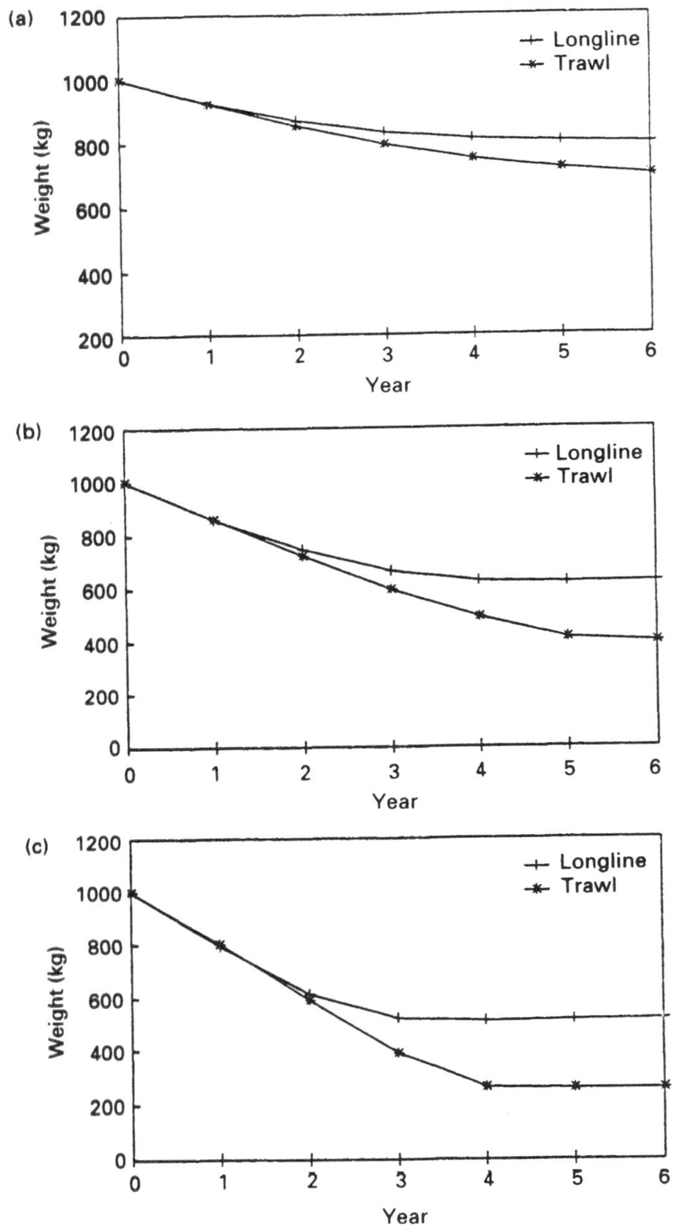

Fig. 7.2 Numerical simulation of the expected biomass reduction of a cod stock (from original 1000 kg), caused by longlining or trawling with three different annual catch levels: (a) 80 kg, (b) 160 kg, and (c) 240 kg. (Adapted from Bjordal & Laevastu, 1990.)

model was developed to simulate the effects on a cod stock harvested by either trawl or longline. For simplicity the model was based on an initial stock biomass of 1000 kg. With increased fishing effort the model as expected predicts a decrease in stock size to gradually lower equilibrium levels after some years of fishing. However, with the same effort or catch level the equilibrium is reached earlier and at a significantly higher stock size with longline fishing than with trawling. In other words, in comparison with trawling, a longline-type exploitation pattern would give a higher long-term yield from a cod stock. The biological effects of different exploitation patterns in specific fisheries should therefore be investigated and potential shifts in effort between various gears involved should be considered in fisheries management strategies in order to improve the yields of the actual stock.

Although the biological effects of different harvesting methods should be of primary and basic importance in fisheries management strategies, the bottom-line result for the fisherman is after all the earnings and profitability of the fishery. Considerations of the biological implications of different management strategies should therefore be extended to bio-economical analyses of different harvesting patterns in multigear fisheries. Again, if trawling and longlining are compared, the longline catches tend to command higher prices as they contain comparatively more large-sized fish which as a rule fetch more than smaller fish. The generally higher efficiency of trawling means that a trawler will normally obtain a considerably higher daily catch and gross income as compared with a similarly sized longline vessel and crew. However, the costs of trawling are considerably higher than for longlining. The basic investment in a trawler is generally higher than for a longline vessel and the costs of operation, particularly the relative fuel costs, are also higher (see Table 7.4). This implies simply that the break-even level in trawl profitability is higher than in longlining, or in other words that relatively more fish have to be caught to make trawl fishing profitable. In general, a certain surplus production from a resource or stock can therefore support a higher profitable employment level with longline fishing than with trawling.

Most of the period after the Second World War has been characterised by technological developments towards higher catching efficiency and a decreasing number of fishermen being able to land increasing amounts of fish. This trend was welcome and generally applauded in most fisheries. Although this development has generally led to increased capture efficiency with most fish capture techniques, longline fishing is still regarded as a relatively labour-intensive fishing method. However, as world catches reached their peak in the 1980s and a large number of major fish stocks were characterised as overexploited, the focus was gradually turned towards conservation-oriented or responsible fishing as well as employment-oriented rather than effective fishing methods. In this context, suggestions have even been made that the fishing efficiency should be decreased by readopting the level of the capture technology used, say, 40 years ago. However, this is not a feasible solution to the problem. Future management strategies should encourage the extended use of fishing techniques that most successfully meet the combined objectives of conservation-oriented fishing, profitability and a high, unsubsidised employment level among fishermen. Although longlining is one of our oldest fishing methods, the different comparative examples given in this chapter suggest that, if evaluated according to the objectives given above, modern longline fishing in the future will be a highly competitive fishing method in a variety of multigear fisheries.

References and Selected Literature

The following list of references and selected literature is not a complete list of the existing literature on longline fishing and related subjects. However, it gives a fairly comprehensive overview of the literature within this field and should provide a good basis for further reading on the different aspects of longline fishing.

Aasen, O. (1963) Line-bight tables – the fishing depth of pelagic longlines. *Fisken og Havet*, **3** (1963), 64 pp (in Norwegian).

Aasen, O. (1965) Hook selection in the longline fishery for spiny dogfish. *Fiskets Gang (Directorate of Fisheries, Norway)*, **1**, 31–33 (in Norwegian).

Adron, J.W. & Mackie, A.M. (1978) Studies on the chemical nature of feeding stimulants for rainbow trout, *Salmo gairdneri* Richardson. *J. Fish. Biol.*, **12**, 303–310.

Agustsson, A. & Ragnarsson, E. (1980) Fuel consumption of the Icelandic fishing fleet. *ICES CM 1980 B*, **5**, 8 pp.

Aldebert, Y., Recasens, L. & Lleonart, J. (1993) Analysis of gear interactions in a hake fishery: the case of the Gulf of Lions (NW Mediterranean). *Sci. Mar.*, **57**(2–3), 207–217.

Allen, K.R. (1963) The influence of behaviour on the capture of fishes with baits. In: *The Selectivity of Fishing Gear*, pp. 5–7. Int. Comm. Northwest Atl. Fish., Spec. Publ. 5, Dartmouth, NS, Canada.

Alverson, D.L., Freeberg, M.H., Murawski, S.A. & Pope, J. (1994) A global assessment of fisheries bycatch and discards. *FAO Fisheries Technical Paper*, FAO, Rome.

Anonymous (1993) *Annual catch statistics of Taiwanese tuna longline fishery 1992.* Tuna Research Center, Institute of Oceanography, National Taiwan University, Taipei, Taiwan.

Arimoto, T. (1984) Interspecific competition among demersal fishes in catch distribution of coastal set-line. *Bull. Jpn. Soc. Sci. Fish*, **50**, 205–210.

Atema, J. (1971) Structures and functions of the sense of taste in the catfish (*Ictalurus natalis*). *Brain Behav. Evol.*, **4**, 273–294.

Atema, J. (1980) Chemical senses, chemical signals and feeding behavior in fishes. In: *Fish Behavior and Its Use in the Capture and Culture of Fishes* (eds J.E. Bardach, J.J. Magnuson, R.C. May & J.M. Reinhart), pp. 57–101. ICLARM Conf. Proc. 5, Manila.

Atema, J. (1988) Distribution of chemical stimuli. In: *Sensory Biology of Aquatic Animals*, pp. 29–56 (eds J. Atema, A.N. Popper, R.R. Fay & W.N. Travolga). Springer-Verlag, New York.

Bardach, J.E. & Atema, J. (1971) The sense of taste in fishes. In: *Handbook of Sensory Physiology*, Vol. IV, pp. 293–336 (ed. L.M. Beidler). Springer-Verlag, Berlin.

Bardach, J.E. & Villars, T. (1974) The chemical senses of fishes. In: *Chemoreception in Marine Organisms*, pp. 49–104 (eds P.T. Grant & A.M. Mackie). Academic Press, New York.

Bertrand, J. (1988) Selectivity of hooks in the handline fishery of the Saya de Malha banks (Indian Ocean). *Fish Res.*, **6**, 249–255.

Beukema, J.J. (1970) Acquired hook-avoidance in the pike *Esox lucius* L. fished with artificial and natural baits. *J. Fish. Biol.*, **2**, 155–160.

Bjordal, Å. (1981) Engineering and fish reaction aspects of longlining – a review. *ICES CM 1981 B*, 35.

Bjordal, Å. (1982) Longline fishery in the future: mechanization, synthetic bait and new hook types. *FTFI-Nytt 4* (in Norwegian) Translated as *Can. Transl. Fish. Aquat. Sci.*, 4978.

Bjordal, Å. (1983a) Full scale trials with new hook designs and reduced size of bait. *ICES CM 1983 B*, 32.

Bjordal, Å. (1983b) Effect of different long-line baits (mackerel, squid) on catch rates and selectivity for tusk and ling. *ICES CM 1983 B*, 31.

Bjordal, Å. (1983c) Mechanized longline fishery in the coastal fishing fleet. Part I. Existing baiting and line handling systems. *FTFI-Rep. 661.8-2-1* (in Norwegian). Translated as: *Can. Transl. Fish. Aquat. Sci.*, 4982, 39 pp.

Bjordal, Å. (1984) The effect of gangion floats on bait loss and catch rates in longlining. *ICES CM 1984 B*, 8.

Bjordal, Å. (1985) Full scale tests of improved longline gear. *ICES CM 1985 B*, 45.

Bjordal, Å. (1986) The behaviour of Norway lobster towards baited creels and size selectivity of creels and trawl. *Fisk. Dir. Skr. Ser. HavUnders.*, **18**, 131–137.

Bjordal, Å. (1987) Swivel connected gangions in mechanized longlining – effect on catch rates and operation. *ICS CM 1987 B*, 18.

Bjordal, Å. (1989) Recent developments in longline fishing – catching performance and conservation aspects. *Proc. World Symp. Fish. Gear Fish Vessel Design, St John's*, Nov. 1988, pp. 19–24.

Bjordal, Å & Laevastu, T. (1990) Effects of trawling and longlining on the yield and biomass of cod stocks – numerically simulated. *ICES CM 1990 G*, 32, Ref. B., 31 pp.

Bjørkum, I. (1993) Technological development of oceanic longlining. *Rep. Norw. Mar. Techn. Inst.* MT40 A93-0105, 402901.00.01, 34 pp (in Norwegian).

Blaxter, J.H.S. (1980) Vision and the feeding of fishes. In: *Fish Behavior and Its Use in the Capture and Culture of Fishes.* (eds J.E. Bardach, J.J. Magnuson, R.C. May & J.M. Reinhart), pp. 32–56. ICLARM Conf. Proc. 5, Manila.

Bowerman, M. (1984) The ancient circle hook re-discovered. *Australian Fisheries*, **43**(6), 34–35.

Brandt, A. von (1984) *Fish Catching Methods of the World.* Fishing News Books, Farnham.

Brawn, V.M. (1969) Feeding behaviour of cod (*Gadus morhua*). *J. Fish. Res. Board Can.*, **26**, 583–596.

Brock, V.E. (1962) On the nature of the selective fishing action of longline gear. *Pacific Science*, **16**, 3–14.

Carr, W.E.S. (1982) Chemical stimulation of feeding behavior. In: *Chemoreception in Fishes*, pp. 259–273 (ed. T.J. Hara). Elsevier, Amsterdam.

Carr, W.E.S. & Derby C.D. (1986) Chemically stimulated feeding behavior in marine animals. *J. Chem. Ecol.*, **12**, 989–1011.

Cornide de Saavedra, J. (1788) Ensayo de una historia de los peces y ostras producciones marinas de la costa de Galicia arreglado al sistema del caballero Carlos Linneo, con un tratado de las diversas pescas, y las redes y aparejos con que se practican. *Ed. en la oficina de Benito Cano*, 264 pp.

Cortez-Zaragoza, E., Dalzell, P. & Pauly, P. (1989) Hook selectivity of yellowfin tuna (*Thunnus albacares*) caught off Darigayos Cove, La Union, Philippines. *J. Appl. Ichthyol.*, **1**, 12–17.

Cross, J.N. (1988) Aspects of the biology of two scyliorhinid sharks, *Apristurus brunneus* and *Parmaturus xaniurus*, from the upper continental slope off Southern California. *Fish. Bull.*, **86**, 691–702.

Daniel, P.C. & Bayer, R.C. (1987) Temporal changes in release rates and quality of lobster (*Homarus americanus*) feeding attractants from herring (*Clupea harengus*) baits. *Mar. Behav. Physiol.*, **13**, 13–27.

Daniel, P.C. & Bayer, R.C. (1989) Fish byproducts as chemo-attractant substrates for the American lobster (*Homarus americanus*): concentration, quality and release characteristics. *Fish. Res.*, **7**, 367–383.

Døving, K.B. & Selset, R. (1980) Behavior patterns in cod released by electrical stimulation of olfactory tract bundlets. *Science*, **207**, 559–560.

Eggers, D.M., Rickard, N.A., Chapman, D.G. & Whitney, R.R. (1982) A methodology for estimating area fished for baited hooks and traps along a ground line. *Can. J. Fish. Aquat. Sci.*, **39**, 448–453.

Ellingsen, O.F. & Døving, K.B. (1986) Chemical fractionation of shrimp extracts inducing bottom food search behavior in cod (*Gadus morhua* L.). *J. Chem. Ecol.*, **12**, 155–168.

Endal, A. (1979) Energy consumption in various Norwegian fisheries. *ICES Fish. Capt. Comm. WG, Gothenburg*, 1979, 7 pp.

Engås, A., Løkkeborg, S., Soldal, A.V. & Ona, E. (1996) Comparative fishing for cod and haddock with commercial trawl and longline at two different stock levels. *J. Northw. Atl. Fish. Sci.* (in press).

Eriksen, T. (1981) *Two Shillings for a Song*. Universitetsforlaget, Tromsø, 281 pp (in Norwegian).

Fernö, A. & Huse, I. (1983) The effect of experience on the behaviour of cod (*Gadus morhua* L.) towards a baited hook. *Fish. Res.*, **2**, 19–28.

Fernö, A., Johannessen, A., Olsen, K.L., Solemdal, P., Tilseth, S. & Valdemarsen, J.W. (1976) Fish behaviour in relation to long lines observed by TV. *ICES CM 1976 B*, 38.

Fernö, A., Solemdal, P. & Tilseth, S. (1986) Field studies on the behaviour of whiting (*Gadus merlangus* L.) towards baited hooks. *FiskDir. Skr. Ser. HavUnders.*, **18**, 83–95.

Forster, G.R. (1973) Line fishing on the continental slope. The selective effect of different hook patterns. *J. Mar. Biol. Assoc. UK*, **53**, 749–751.

Franco, J.M. Bjordal, Å. & Løkkeborg, S. (1987) Synergistic effects of different baits on the same hook in the longline fishery for hake (*Merluccius merluccius*) and tusk (*Brosme brosme*). *ICES CM 1987 B*, 39.

French, R.R. (1969) Comparison of catches of Pacific salmon by gill nets, purse seines and longlines. *Bull. Int. N. Pacif. Fish. Comm.*, **26**, 13–26.

Goh, Y. & Tamura, T. (1979) Olfactory responses to amino acids in marine teleosts. *Comp. Biochem. Physiol.*, **62A**, 863–868.

Gong, Y., Lee, J.U., Kim, Y.S. & Yang, W.S. (1989) Fishing efficiency of Korean regular and deep longline gears and vertical distribution of tunas in the Indian Ocean. *Bull. Korean Fish. Soc.*, **22**, 86–94.

Gong, Y., Cho, K.D., Yang, W.S., Park, Y.C. & Kim, T.I. (1995) Abundance of tunas by longliners in relation to oceanographic conditions in the tropical Pacific Ocean. *Bull. Nat. Fish. Res. Dev. Agency* **49**, 37–47.

Grimes, C.B., Able, K.W. & Turner, S.C. (1982) Direct observation from a submersible vessel of commercial longlines for tilefish. *Trans. Am. Fish. Soc.*, **111**, 94–98.

Halliday, R.G. & Clark, K.J. (1995) The Scotia–Fundy region groundfish hook and line fisheries: a digest of quantities and sizes landed, and comparison with other gear types. Canadian Manuscript Report of Fisheries and Aquatic Science 2271, Biological Sciences Branch, Department of Fisheries and Oceans, Bedford Institute of Oceanography, Dartmouth, Nova Scotia, Canada.

Halliday, R.G. & Kenchington, T. (1993) Size selection of groundfish longline gear. Project Summary, Fisheries and Oceans, Scotia–Fundy Region, Halifax, Nova Scotia. No. 40, April 1993, 5 pp.

Hamley, J.M. & Skud, B.E. (1978) Factors affecting longline catch and effort: II. Hook-spacing. *Int. Pac. Halib. Comm. Sci. Rep.*, **65**, 15–24.

Hamre, J. (1968) The relative catching efficiency of a new hook design in the longline fishery for spiny dogfish. *Fiskets Gang (Directorate of Fisheries, Norway)*, **46**, 793–796 (in Norwegian).

Hara, T.J. (1975) Olfaction in fish. In: *Progress in Neurobiology, Vol. 5*, (eds G.A. Kerkut & J.W. Phillis), pp. 271–335. Pergamon Press, Oxford.

Hara, T.J. (1993) Role of olfaction in fish behaviour. In: *Behaviour of Teleost Fishes*, (ed. T.J. Pitcher), pp. 171–199. Chapman & Hall, London.

Hart, P.J.B. (1986) Foraging in teleost fishes. In: *The Behaviour of Teleost Fishes*, (ed. T.J. Pitcher), pp. 211–235. Croom Helm, London.

Hearn, D. & Warren, J. (1980) An assessment of monofilament lines in the Newfoundland longline fishery. *Rep. Dept. Fish., St. John's, Newfoundland*, 12 pp.

High, W.L. (1980) Bait loss from halibut longline gear observed from submersible. *Mar. Fish. Rev.*, **42**, 26–29.

Hirayama, B. (1969) Studies on the fishing mechanism of tuna longline – III. The difference of catch by retrieving methods. *Bull. Jpn. Soc. Sci. Fish.*, **35**, 629–635.

Hovgård, H. & Riget, F.F. (1992) Comparison of long-line and trawl selectivity in cod surveys off West Greenland. *Fish. Res.*, **13**, 323–333.

Hurum, H.J. (1977) *A History of the Fish Hook*. Adam & Charles Black, London.

Huse, I. (1979) *Influence of hook design and gear materials in longlining for cod (*Gadus morhua *L.) and haddock (*Melanogrammus aeglefinus *L.) inves-*

tigated by behaviour studies and fishing trials. Cand. real.-thesis, University of Bergen, Bergen, Norway, 122 p (in Norwegian).

Huse, I. (1979) New hook designs in the longline fisheries for cod, haddock, ling and tusk. *Rep. Inst. Fish. Techn. Res.* (661. 5-1-1), 14 pp (in Norwegian).

Huse, I. & Fernö, A. (1990) Fish behaviour studies as an aid to improved longline hook design. *Fish Res.*, **9**, 287–297.

Imai, T. (1972) Studies on the several raw fish baits in tuna long-line. Fishing I. *Mem. Fac. Fish., Kagoshima Univ.*, **21**, 45–50.

Imai, T. & Shirakawa, O. (1972) Studies on the several raw fish baits in tuna longline. Fishing II. *Mem. Fac. Fish., Kagoshima Univ.*, **21**, 51–62.

Japp, D.W. (1993) Longlining in South Africa. In: *Fish, Fishers and Fisheries – Proceedings of the Second South African Marine Linefish Symposium*, Durban, 23–24 October, Oceanographic Research Institute, Special Publication No. 2, pp. 134–139.

Johannes, R.E. (1981) *Words of the Lagoon – Fishing and Marine Lore in the District of Micronesia*. University of California Press, Los Angeles.

Johannes, R. (1984) Island study shows why circle hooks work so well. *Australian Fisheries*, Oct. 1984, 34–36.

Johannessen, T. (1983) *Influence of hook and bait size on catch efficiency and length selection in longlining for cod (*Gadus morhua *L.) and haddock (*Melanogrammus aeglefinus *L.)* Cand. real.-thesis, University of Bergen, Bergen, Norway, 109 pp (in Norwegian).

Johannessen, T. (1984) Fishing experiments with different longline baits for cod, February 1984. *ICS Ad Hoc Working Group on Artificial Bait and Bait Attractions*, Hirshals, 1984 [mimeo].

Johannessen, T., Fernö, A. & Løkkeborg, S. (1993) Behaviour of cod (*Gadus morhua*) and haddock (*Melanogrammus aeglefinus*) in relation to various sizes of long-line bait. *ICES Mar. Sci. Symp.*, **196**, 47–50.

Johnstone, A.D.F. (1980) The detection of dissolved amino acids by the Atlantic cod, *Gadus morhua* L. *J. Fish. Biol.*, **17**, 219–230.

Johnstone, A.D.F. & Hawkins, A.D. (1981) A method for testing the effectiveness of different fishing baits in the sea. *Scottish Fisheries Information Pamphlet 3*, 7 pp.

Johnstone, A.D.F. & Mackie, A.M (1984) Investigations of bait acceptance by the cod, *Gadus morhua* L. *ICES CM 1984 B*, 28, 13 pp.

Jones, K.A. (1992) Food search behaviour in fish and the use of chemical lures in commercial and sports fishing. In: *Fish Chemoreception* (ed. T.J. Hara), pp. 288–320. Chapman & Hall, London.

Kaimmer, S.M. (1994) Halibut injury and mortality associated with manual and automated removal from setline hooks. *Fish. Res.*, **20**, 165–179.

Kanda, K., Koike, A. Takeuchi, S. & Ogura, M. (1978) Selectivity of the hook for mackerel, *Scomber japonicus houttuyn*, pole fishing. *J. Tokyo Univ. Fish.*, **64**, 109–114.

Karlsen, L. (1976) Basic studies of Norwegian longline gear. *ICES CM 1976 B*, 27.

Karlsen, L. (1989) *Fishing Gear Technology*. Universitetsforlaget Oslo. 207 pp (in Norwegian).

Kenchington, T.J. (1991) Some effects of bottom trawling on the availability of cod biomass to an inshore longline fishery: a discussion. *NAFO SCR. Doc. 91/101, Serial No. N1993.*

Kenchington, T.J. (1993) Estimation of the size selectivity of baited hooks: a graphical approach. *Can. J. Fish. Aquat. Sci.*, **50**, 772–780.

Kenchington, T.J. & Halliday, R.G. (1994) A survey of fishing practices in the Scotia–Fundy region groundfish longline fisheries. *Can. Manuscr. Rep. Fish. Aquat. Sci.*, 2225, 642 pp.

Kenchington, T.J., Halliday, R.G. & Harrison, G.D. (1994) Fishing grounds exploited in 1990 by groundfish longliners in Canada's Scotia–Fundy region. *NAFO Sci. Counc. Studies* 20.

Kleerekoper, H. (1982) The role of olfaction in the orientation of fishes. In: *Chemoreception in Fishes* (ed. T.J. Hara), pp. 201–225. Elsevier, Amsterdam.

Kleerekoper, H., Gruber, D. & Matis, J. (1975) Accuracy of localization of a chemical stimulus in flowing and stagnant water by the nurse shark. *Ginglymostoma cirratum. J. Comp. Physiol.*, **98**, 257–275.

Klein, S.J. (1986) Selectivity of trawl, trap, longline and set-net gears to sablefish, *Anoplopoma fimbria.* US Dept of Commerce, Nat. Mar. Fish. Ser., Alaska Fisheries Center, 84 pp.

Ko, K.S. & Kim, Y.H. (1981) The deformation and breaking load of the fishing hook by the tensile test. *Bull. Korean Fish. Soc.*, **14**(4), 269–275.

Ko, K.S. & Kim, Y.H. (1982) The jerking force by the hooked carp and its periodicity with the tail beat. *Bull. Korean Fish. Soc.*, **15**(3), 226–232.

Kobayashi, H. (1975) A study on the success of a fishery with artificial baits for tuna long lines. *Bull. Jpn. Soc. Sci. Fish.*, **41**, 175–182.

Koike, A. & Kanda, K. (1978) Selectivity of the hook of pond smelt, *Hypomesus olidus. J. Tokyo Univ. Fish.*, **64**, 115–123.

Koike, A., Takeuchi, S., Ogura, M., Kanda, K. & Arihara, C. (1968) Selection curve of the hook of long line. *J. Tokyo Univ. Fish.*, **55**, 77–82.

Kurogane, K. (1968) Experimental comparison of fishing power of longline for bottom fishes in the North Pacific. *Bull. Tokai reg. Fish. Res. Lab.*, **55**, 115–128.

Lee, C., Igarashi, S., Mikami, T. and Yamashita, N. (1989) Behaviour studies of walleye pollock *Theragra chalcogramma* to a bait and baited hook for hook design. *Nippon Suisan Gakkaishi*, **55**, 1553–1558.

Lee, C., Igarashi, S., Mikami, T. & Yamashita, N. (1990) A hydrodynamical analysis of hooking process of walleye pollock *Theragra chalcogramma. Nippon Suisan Gakkaishi*, **56**, 43–49.

Lleonart, J. & Camarasa, J.M. (1987) *La pesca a Catalunya el 1722.* Estudis Maritims Vol 1, Museu Maritim, Diputacio Barcelona, 127 pp.

Løkkeborg, S. (1985) *Factors that influence the catchability of natural and artificial bait.* M.Sc. thesis, University of Bergen, Norway, 118 pp (in Norwegian).

Løkkeborg, S. (1989) *Longline bait: fish behaviour and the influence of attractant release rate and bait appearance.* Dr. Scient.-thesis, University of Bergen, Norway, 109 pp.

Løkkeborg, S. (1990a) Rate of release of potential feeding attractants from natural and artificial bait. *Fish. Res.*, **8**, 253–261.

Løkkeborg, S. (1990b) Reduced catch of under-sized cod (*Gadus morhua*) in longlining by using artificial bait. *Can. J. Fish. Aquat. Sci.*, **47**, 1112–1115.

Løkkeborg, S. (1990c) Fish shaped bait does not improve the response of coastal cod. *ICES CM 1990 B*, 39.

Løkkeborg, S. (1991) Fishing experiments with an alternative longline bait using surplus fish products. *Fish. Res.*, **12**, 43–56.

Løkkeborg, S. & Bjordal, Å. (1992a) Species and size selectivity in longline fishing: a review. *Fish Res.*, **13**, 311–322.

Løkkeborg, S. & Bjordal, Å. (1992b) Reduced bait loss and bycatch of seabirds in longlining by using a seabird scarer. *Document to Working Group FSA-92 CCAMLR*, Hobart, Australia.

Løkkeborg, S. & Bjordal, Å. (1995) Size-selective effects of increasing bait size by using an inedible body on longline hooks. *Fish. Res.*, **24**, 273–279.

Løkkeborg, S. & Johannessen, T. (1992) The importance of chemical stimuli in bait fishing – fishing trials with presoaked bait. *Fish. Res.*, **14**, 21–29.

Løkkeborg, S., Bjordal, Å. & Fernö, A. (1989) Responses of cod (*Gadus morhua*) and haddock (*Melanogrammus aeglefinus*) to baited hooks in the natural environment. *Can. J. Fish. Aquat. Sci.*, **46**, 1478–1483.

Løkkeborg, S. Bjordal, Å. & Fernö, A. (1993) The reliability and value of studies of fish behaviour in long-line gear research. *ICES Mar. Sci. Symp. 196*, 41–46.

Løkkeborg, S., Olla, B.L., Pearson, W.H. & Davis, M.W. (1995) Behavioural responses in sablefish, *Anoplopoma fimbria*, to bait odour. *J. Fish Biol.*, **46**, 142–155.

Mackie, A.M. (1982) Identification of the gustatory feeding stimulants. In: *Chemoreception in Fishes* (ed. T.J. Hara), pp. 275–291. Elsevier, Amsterdam.

Mackie, A.M. & Adron, J.W. (1978) Identification of inosine and inosine-5′-monophosphate as the gustatory feeding stimulants for the turbot, *Scophthalmus maximus*. *Comp. Biochem. Physiol.*, **60A**, 79–83.

Maeda, H. (1960) A tentative analysis of the distribution pattern of tuna projected on the longline. *J. Shimoneseki College Fish*, **9**, 89–308.

Maeda, H., Kitajima, T., Hirose, M., Uchida, K., Hamaguchi, M. & Tawara, S. (1982) Mechanism of bottom longline to catch groundfish – consideration on interspecific competition among groundfishes observed in baited hook occupation. *Bull. Sekai Reg. Fish. Res. Lab. No. 57* (1982).

Makiguchi, N., Arita, M. & Asai, Y. (1980) Application of a luminous bacterium to fish-attracting purpose. *Bull. Jpn. Soc. Sci. Fish.*, **46**, 1307–1312.

March, E.J. (1953) *Sailing Trawlers – the Story of Deep-sea Fishing with Long Line and Trawl.* Percival Marshall and Co. Ltd, London.

Martin, W.R. & McCracken, F.D. (1954) Relative efficacy of baits for groundfish. *Prog. Rep. Atl. Biol. Sta.*, **126**, 17–20.

McCracken, F.D. (1954) Hook selection trials. *Fish. Res. Bd Can., Rept Atl. Biol. Stat. 1954*, 57–60.

McCracken, F.D. (1963) Selection by codend meshes and hooks on cod, haddock, flatfish and redfish. In: *The Selectivity of Fishing Gear.* Int. Comm. Northwest Atl. Fish., Spec. Publ. 5, Dartmouth, NS, Canada, pp. 137–155.

McEachron, L.W., Doerzbacher, J.F., Matlock, G.C., Green, A.W. & Saul, G.E. (1988) Reducing the bycatch in a commercial trotline fishery. *Fish. Bull.*, **86**, 109–117.

McGregor, I.R. (Ed.) (1994) *Report of Assessment and Research Activities 1993.* International Pacific Halibut Commission, Seattle, Washington, USA.

Myhre, R.J. (1969) Gear selection and Pacific halibut. *IPHC Sci. Rept 51*, 35 pp.
Myhre, R.J. & Quinn, T.J. (1984) Comparison of efficiency of snap gear to fixed-hook setline gear for catching Pacific halibut. *IPHC Sci. Rep. 69*, 37 pp.

Nedreaas, K., Soldal, A.V. & Bjordal, Å. (1993) Performance and biological implications of a multi-gear fishery for Greenland halibut (*Reinhardtius hippoglossoides*). *NAFO Sci. Doc. 93/118, Serial No. N2312.*
Nedreaas, K., Sandberg, P. & Veim, A.K. (1995) Biological and economic implications of a multi-gear fishery for Greenland halibut (*Reinhardtius hippoglossoides*). *ICES CM 1995 S*, 5, 25 pp.
Neilson, J.D., Waiwood, K.G. & Smith, S.J. (1989) Survival of Atlantic halibut (*Hippoglossus hippoglossus*) caught by longline and otter trawl gear. *Can. J. Fish. Aquat. Sci.*, **46**, 887–897.
Nygaard, W. (1988) *A Guide to Longlining.* O. Mustad & Son Ltd, Gjøvik, Norway.

Ohshima, Y. (1953) On the pull of fish caught by fish-hook. *Bull. Jpn. Soc. Sci. Fish.*, **19**, 233–238.
Olsen, S. & Laevastu, T. (1983a) Fish attraction to baits and effects of currents on the distribution of smell from baits. *Proc. Rep. Northwest and Alaska Fisheries Center, Seattle*, 45 pp.
Olsen, S. & Laevastu, T. (1983b) Factors affecting catch of long lines, evaluated with a simulation model of longline fishing. *Proc. Rep. Northwest and Alaska Fisheries Center, Seattle*, 50 pp.
Otway, N.M. & Craig, J.R. (1993) Effects of hook size on the catches of under-sized snapper *Pagrus auratus*. *Mar. Ecol. Prog. Ser.*, **93**, 9–15.

Park, S.W. (1972) The fishing efficiency of the tuna longline. 1: The distribution of the hooking depth and hooking temperature. *Bull. Pusan Fish. Coll.*, **12**(1), 16 pp.
Park, S.W. (1974a) The fishing efficiency of the tuna longline. 2: The fishing efficiency decrease by soaking time. *Bull. Pusan Fish. Coll.*, **14**(1), 11 pp.
Park, S.W. (1974b) Fishing power and diurnal variations of catchability of tuna longline. *Publ. Mar. Lab. Busan Fish. Coll.*, **7**, 51–76.
Park, S.W. (1982a) An analysis of the frequencies of the saury heads (bait) retained in the tuna stomachs. *Bull. Korean Fish. Soc.*, **15**(4), 312–316.
Park, S.W. (1982b) The biting behaviour of tuna on baits. *Bull. Korean Fish. Soc.*, **15**(4), 317–322.
Pawson, M.G. (1977a) The responses of cod *Gadus morhua* (L) to chemical attractants in moving water. *J. Cons. Int. Explor. Mer.*, **37**(3), 316–318.
Pawson, M.G. (1977b) Analysis of a natural chemical attractant for whiting *Merlangius merlangus* L. and cod *Gadus morhua* L. using a behavioural bioassay. *Comp. Biochem. Physiol.*, **56A**, 129–135.
Pearson, W.H., Miller, S.E. & Olla, B.L. (1980) Chemoreception in the food-searching and feeding behavior of the red hake, *Urophycis chuss* (Walbaum). *J. Exp. Mar. Biol. Ecol.*, **8**, 139–150.
Peeling, D. & Rodgers, D.M. (1985) Circle hook comparison study. *Fisheries and Oceans, Scotia–Fundy region, Halifax, Canada, Project Rep. No 78*, 16 pp.
Punt, A.E. (1994) Assessment of the stocks of Cape hakes *Merluccius* spp. off South Africa. *S. Afr. J. mar. Sci.*, **14**, 159–186.

Ralston, S. (1982) Influence of hook size in the Hawaiian deep-sea handline fishery. *Can. J. Fish. Aquat. Sci.*, **39**, 1297–1302.

Ralston, S. (1990) Size selection of snappers (*Lutjanidae*) by hook and line gear. *Can. J. Fish. Aquat. Sci.*, **47**, 696–700.

Rollefsen, G. (1953) The selectivity of different fishing gear used in Lofoten. *J. de Conceil*, **19**, 191–194.

Rothschild, B.J. (1967) Competition for gear in a multiple-species fishery. *J. de Conceil*, **31**(1), 102–110.

Russel, G.M., Gutherz, E.J. & Barans, C.A. (1988) Evaluation of demersal longline gear off South Carolina and Puerto Rico with emphasis on deep-water reef fish stocks. *Mar. Fish. Rev.*, **50**(1), 26–31.

Sætersdal, G. (1963) Selectivity of longlines. *Spec. Publ. Int. Comm. NW Atlant. Fish.*, (5), 189–192.

Sainsbury, J.C. (1986) *Commercial Fishing Methods.* Fishing News Books, Farnham Surrey.

Sakagawa, G.T., Coan, A.L. & Bartoo, N.T. (1987) Patterns in longline fishery data and catches of bigeye tuna, *Thunnus obesus. Mar. Fish. Rev.*, **49**, 57–66.

Sanez Reguart, A. (1791–1795) *Diccionario Historico de los Artes de Pesca Nacional.* Imprenta de la Viuda de Don Joaquin Ibarra, Madrid, V vols.

Shepard, M.P., Roberts, R.F.A., Aro, K.V. & Turner, C.E. (1975) Effect of bait loss on catching power of floating longline gear. *Bull. Int. North Pac. Fish. Comm.*, **32**, 71–75.

Shimada, K. (1972) On the bait for tuna long line – III. N.T. fishing light type L.L. of acryl resin. *Mem. Fac. Fish., Kagoshima Univ.*, **21**(1), 79–89.

Shimada, K. & Tsurudome, M. (1971) On the bait for tuna long-line – II. On the saury, mackerel and mackerel scad baits for tuna fishing. *Mem. Fac. Fish., Kagoshima Univ.*, **20**(1), 119–130.

Shimozaki, Y., Fujii, K. & Taketomi, K. (1983) On the coefficient of friction, and the frictional force between a main-line and the rollers of a line hauler. *Bull. Natl Res. Inst. Fish. Eng.*, **4**, 221–233.

Sigler, M.F. (1993) *Stock assessment and management of sablefish,* Anoplopoma fimbria, *in the Gulf of Alaska.* PhD Thesis, University of Washington, Washington, USA.

Sigler, M.F. & Fujioka, J.T. (1988) Evaluation of variability in sablefish, *Anoplopoma fimbria*, abundance indices in the Gulf of Alaska using the bootstrap method. *Fish. Bull.*, **86**(3), 445–452.

Sinclair, A.F. (1986) Longliner–otter trawler interactions in cod fisheries on the Scotian Shelf: implications of differences in partial recruitment. *CAFSAC Res. Doc.*, **86**, 94.

Sivasubramaniam, K. (1961) Relation between soaking time and catch of tunas in longline fisheries. *Bull. Jpn. Soc. Sci. Fish.*, **27**, 835–845.

Skeide, R., Bjordal, P. & Løkkeborg, S. (1986) Testing of a new hook design (E-Z-baiter) through comparative longline fishing trials. *ICES CM 1986 B*, 25.

Skud, B.E. (1978) Factors affecting longline catch and effort: I General review. *IPHC Sci. Rept*, **64**, 5–14.

Skud, B.E. (1978) Factors affecting longline catch and effort: III. Bait loss and competition. *IPHC Sci. Rept*, **64**, 25–43.

Sutterlin, A.M. (1975) Chemical attraction of some marine fish in their natural habitat. *J. Fish. Res. Bd Can.*, **32**, 729–738.

Sutterlin, A.M., Solemdal, P. & Tilseth, S. (1982) Baits in fisheries with emphasis on the North Atlantic cod fishing industry. In: *Chemoreception in Fishes* (ed. T.J. Hara), pp. 293–305. Elsevier, Amsterdam.

Takagi, K. (1971) Information on the catchable time period for Pacific salmon obtained through simultaneous fishing by longlines and gillnets. *Far Seas Fish. Res. Lab., Shimuzu, Japan. Bull.*, **5**, 177–194.

Takeuchi, S. & Koike, A. (1969) The effect of size and shape of hook on the catching efficiency and selection curve of longline. *J. Tokyo Univ. Fish.*, **55**, 119–124.

Templeman, W. (1967) Predation on living fishes on longline in Baffin Bay by the amphipod *Eurythenes gryllus* (Lichtenstein), and a new distribution record. *J. Fish. Res. Bd Can.*, **24**, 215–217.

Templeman, W. & Fleming, A.M. (1956) The Bonavista longlining experiment, 1950–1953. *Bull. Fish. Res. Bd Can.*, **109**, 55 pp.

Tilseth, S., Solemdal, P. & Fernö, A. (1978) Behaviour studies of fish reaction to long lines. *ICES CM 1978 B*, **22**, 9 pp.

Wilson, R.R. & Smith, K.L. (1984) Effect of near-bottom currents on detection of bait by the abyssal grenadier fishes *Coryphaenoides* spp, recorded in situ with a video camera on a free vehicle. *Mar. Biol.*, **84**, 83–91.

Yamaguchi, Y. & Hidaka, I. (1985) A short review of the recent long-line fishing and developmental works on artificial baits in Japan. *ICES Ad Hoc Working Group on Artificial Bait and Bait Attraction*, Bergen.

Yamaguchi, Y., Nonoda, T., Kobayashi, H., Izawa, K., Jinno, T., Ishikura, I., Uchida, M. & Tonogai, M. (1983) Effectiveness of artificial bait for obtaining higher hooking rate on bottom set long-line fishing. *Bull. Jpn. Soc. Sci. Fish.*, **42**, 1819–1824.

Yamaguchi, Y., Nonoda, T., Hidaka, I., Niwa, E., Jinno, T., Ishikura, I. & Uchida, M. (1984) Ribbon type artificial baits used for bottom set longline fishing. *Bull. Fac. Fish. Mie Univ.*, **11**, 95–109.

Yang, W.S. & Gong, Y. (1988) The vertical distribution of tunas and billfishes, and fishing efficiency between Korean regular and deep longlines in the Atlantic Ocean. *Bull. Natl Fish. Res. Dev. Agency (Korea)*, **49**, 39–42.

Yang, W.S. & Park, Y.C. (1990) Monthly changes of catch per unit effort of bigeye and yellowfish tunas for the Korean and Japanese longline fishery in the Atlantic Ocean, 1984–1986. *Bull. Natl Fish. Res. Dev. Agency*, **44**, 47–57.

Yetman, L., Hogan, L. & Lacey, M. (1991) Longline hook selectivity study: south coast of Newfoundland. *AFAP Newfoundland Region, Gear Conversion, Project Report 1991*, *181*, 19 pp.

Zeiske, E., Theisen, B. & Breuker, H. (1992) Structure, development and evolutionary aspects of the peripheral olfactory system. In: *Fish Chemoreception* (ed. T.J. Hara), pp. 13–39. Chapman & Hall, London.

Glossary

Active space In fishing with baited gears, the volume within which feeding attractants released from the bait are present in concentrations above the chemosensory threshold of the fish.

Active gear Fishing gears that are used to actively pursue the fish, e.g. trawls, seine nets, purse seines and harpoons. (*See* passive gear.)

Alimentary tract The digestive tract from mouth to anus.

Anchor Device normally made from steel with a stem and between two and five claws, used to secure fishing gears such as longlines and gillnets to the bottom.

Artificial bait In contrast to natural bait, artificial bait is a processed product, made either from restructured raw material such as fish or from chemical smell stimuli – in both cases combined with a binder to secure adequate release of stimuli and in the case of longline bait also some reinforcing material to make the bait stay on the book. (*See* bait, natural bait.)

Autoline The brand name of a mechanised longline handling and baiting system manufactured by O. Mustad & Son Ltd. Commonly used as a general term for automated longline systems.

Backline *See* mainline.

Bait Substance releasing feeding stimulants that attract fish and other capture organisms. (*See* natural bait, artificial bait.)

Bait scavenger Non-target organism feeding on the bait.

Balloon *See* buoy.

Barb Projection close to the point of a fish hook, to secure the hook properly in the fish flesh.

Barbel Taste appendage located on the head of a fish, used in searching for food on the sea bed.

Bladder *See* buoy.

Bottom longline Longline set on the bottom (sea bed).

Buoy Common term for a device floating on the water. In longlining the buoy is connected to the buoy line and anchor, and is thus a part of the system for gear retrieval. Today commonly made from plastic, inflatable with a close to spherical or bulbous shape.

Buoy line Line (rope) connecting buoy and anchor, used to retrieve the gear.

Bycatch Catch of commercial value other than the target species.

Catchability The catching power (efficiency) of a fishing gear, defined as the proportion of available fish taken by the gear.

Chemotaxis Behavioural response to chemical stimuli.

Chute Metal construction used to shield the longline during setting from tubs, to prevent hooks from fastening in the railing or other parts of the vessel.

Demersal fish Fish inhabiting the water column close to the sea bed during most of their life cycle (e.g. cod, haddock, halibut, ling, sablefish, tusk).

Demersal longline *See* bottom longline.

Detection threshold The lowest strength of a stimulus (e.g. light intensity, sound level, concentration of chemical compounds) that the sensory organ of a fish is capable of detecting.

Discard Catch of no commercial value, which is discarded at sea.

Drum gear Longlining where the mainline is spooled on drums and the hooks and snoods are attached/detached during setting and hauling, respectively.

Feeding attractant Stimulus, released from food items, to which an animal responds by orienting towards the source, possibly over long distances.

Feeding stimulant Stimulus, released from food items, that promotes ingestion and continuation of feeding; sometimes used to mean feeding attractant.

Fixed gear *See* stationary gear.

Fleet Unit of longline gear, consisting of several skates linked together. The length of the fleet varies with fishing conditions.

Float Buoyancy device, used either to keep the upper end of anchor lines afloat at the surface or to lift parts of the gear off the bottom. Surface floats are often inflatable plastic bulbs, while submerged floats are usually made from hard plastic, glass or metal to withstand the water pressure. (*See* buoy.)

Gaff hook Steel hook with wooden handle used by the fisherman to grab the fish on the longline and bring it on board.

Gangion *See* snood.

Gear saturation Competition among fish for a unit of gear: in longlining, fish may compete for available baits.

Grapnel *See* anchor.

Groundfish *See* demersal fish.

Groundline *See* mainline.

Gustatory organ/system The taste organ which in fish is composed of receptor cells gathered into structures called taste buds.

Halibut snap *See* snap.

Hook Bent metal device used to catch fish.

Longline Fishing gear consisting of long (main line) to which subsidiary lines (snoods) with baited hooks are attached at intervals.

Mainline Part of longline gear: long rope or monofilament line to which snoods with baited hooks are attached.

Marker buoy Float with pole and flag(s) used with stationary fishing gears, to keep the surface end of the anchor/buoy line afloat as well as being a visible marker for easy spotting and retrieval of the gear.

Monofilament Line made of one (usually polyamide) fibre.

Multifilament Line (rope) made of many thin fibres.

Natural bait Piece of flesh from an organism (or whole organism) used to attract, aggregate and capture fish and other organisms on longlines and handlines and in pots, creels and traps.

Odour plume Feeding attractants released from baits and spread by the water current generate a plume of scent to which fish respond by exhibiting food searching behaviour. (*See* active space.)

Olfactory organ/system The smell organ (nose) which in fish is a paired organ with an olfactory cavity that has an anterior and a posterior nostril.

Passive gear Fishing gear based on capture by active (voluntary) movements of the fish relative to the gear, in contrast to active gears.

Pelagic fish Fish inhabiting open waters (e.g. herring, mackerel, salmon, sardine, tuna).

Pelagic longline Longline gear submerged from surface floats, either anchored or drifting freely.

Precise baiting Controlled automated baiting of longline hooks.

Random baiting Random or uncontrolled baiting of longline hooks.

Release rate The temporal change in the concentrations of attractants released from baits.

Response threshold The lowest strength of a stimulus (e.g. light intensity, sound level, concentration of chemical compounds) to which a fish will respond. This stimulus strength is often significantly higher than the detection threshold.

Rheotaxis Behavioural response to flow or water current.

Selection Species and sizes of fish caught (selected) by a fishing gear relative to the actual species and size composition in an area.

Semipelagic longline Longline anchored or fixed to the bottom by anchors and weights, but lifted off the bottom by floats.

Setting Putting fishing gear into the water.

Shooting *See* setting.

Size selectivity The proportion of each size group of fish present on the fishing ground that is caught and retained by the gear.

Skate Certain unit or length of longline gear, usually used for the length of gear that fits into a tub.

Snap Metal clip (may look like a large safety pin, often with a swivel) that is used to attach the snood and hook to the mainline.

Snap-on gear Longline gear, based on using snaps, e.g. halibut, tuna and swordfish longlines.

Snood Thin line connecting the hook to the mainline of the longline gear.

Soak time The fishing period of stationary fishing gears, often from the time of setting to the time of hauling.

Species selectivity The proportion of each species of fish present on the fishing ground that is caught and retained by the gear.

Stationary gear Non-moving fishing gear, fish capture being based on attraction of fish to the gear or fish/gear encounters by expected or arbitrary fish movements.

Stock assessment Estimating the abundance of fisheries resources.

Swivel Device used to allow free rotation between two linked objects. In longlining used in the linkage between snood and mainline to avoid twisting of the snood.

Target species The one or the few species mainly sought after in a specific fishing operation.

Taste bud *See* gustatory organ.

Threshold *See* detection threshold *and* response threshold.

Total allowable catch (TAC) Total annual catch allowed to be taken for a given fish stock. This figure is set by management bodies based on stock assessment works.

Trawl Today this word is generally used for fishing gears such as otter trawls, beam trawls, pair trawls, etc. Earlier commonly used for a certain length of longline gear, and the term may still be used locally in this sense.

Weight In longlining, weights are used on the line to make it sink faster and to make it lie properly on the sea bed, or in combination with float and float line to fix the longline gear at a certain depth above the bottom.

Common and scientific names of some important longline target species.

Common name	Scientific name
Black scabbard	*Aphanopus carbo*
Cod (Atlantic)	*Gadus morhua*
Cod (Pacific)	*Gadus macrocephalus*
Greenland halibut	*Reinhardtius hippoglossoides*
Haddock	*Melanogrammus aeglefinus*
Hake	*Merluccius* sp.
Halibut (Atlantic)	*Hippoglossus hippoglossus*
Halibut (Pacific)	*Hippoglossus stenolepis*
Kingklip (south west Atlantic)	*Genypterus blacodes*
Kingklip (south east Atlantic)	*Genypterus capensis*
Ling	*Molva molva*
Patagonian toothfish	*Dissostichus eleginoides*
Sablefish	*Anoplopoma fimbria*
Saithe	*Pollachius virens*
Tuna	
Albacore	*Thunnus alalunga*
Bigeye	*Thunnus obesus*
Bluefin	*Thunnus thynnus*
Yellowfin	*Thunnus albacares*
Tusk	*Brosme brosme*
White hake	*Urophycis tenuis*
Wolffish	*Anarhichas* sp.

Index

Books published by **Fishing News Books**

Free catalogue available on request from Fishing News Books, Blackwell Science, Osney Mead, Oxford OX2 0EL, England

Abalone farming
Abalone of the world
Advances in fish science and technology
Aquaculture
Aquaculture and the environment
Aquaculture and water resource management
Aquaculture development – progress and prospects
Aquaculture in Taiwan
Aquaculture: principles and practices
Aquaculture systems
Aquaculture training manual
Aquatic ecology
Aquatic microbiology
Aquatic predators and their prey
Aquatic weed control
Atlantic salmon: its future
The Atlantic salmon: natural history etc.
Bacterial diseases of fish
Better angling with simple science
Bioeconomic analysis of fisheries
Bivalve molluscs
British freshwater fishes
Broodstock management and egg and larval quality
Business management in fisheries and aquaculture
Cage aquaculture
Calculations for fishing gear designs
Carp farming
Carp and pond fish culture
Catch effort sampling strategies
Commercial fishing methods
The common fisheries policy
Control of fish quality
Crab and lobster fishing
The crayfish
Crustacean farming
Design of small fishing vessels
Developments in electric fishing
Developments in fisheries research in Scotland
Dynamics of marine ecosystems
Ecology of fresh waters
The economics of salmon aquaculture
The edible crab and its fishery in British waters
Eel culture
The effects of fishing on marine ecosystems and communities
Engineering, economics and fisheries management
The European fishing handbook 1993–94
Exploitable marine ecosystems: their behaviour and management
FAO catalogue of fishing gear designs
FAO catalogue of small scale fishing gear
Fibre ropes for fishing gear
Fish catching methods of the world
Fisheries biology, assessment and management
Fisheries management in crisis
Fisheries mismanagement – the case of the North Atlantic cod
Fisheries of Australia
Fisheries sonar
Fisherman's workbook
Fishermen's handbook
Fishery development experiences
Fishing and stock fluctuations
Fishing boats and their equipment
Fishing boats of the world 1
Fishing boats of the world 2
Fishing boats of the world 3
Fishing ports and markets
Fishing with electricity
Fishing with light
Fish preservation and processing
Freshwater fisheries management
Fundamentals of aquatic ecology
Glossary of UK fishing gear terms
Handbook of trout and salmon diseases
A history of marine fish culture in Europe and North America
How to make and set nets
The Icelandic fisheries
Inland aquaculture development handbook
Intensive fish farming
Large marine ecosystems – patterns, processes and yields
Large marine ecosystems – stress, mitigation and sustainability
Large marine ecosystem sustainability: the Northeast Shelf
The law of aquaculture: the law relating to the farming of fish and shellfish in Great Britain
A living from lobsters
Longlining
Making and managing a trout lake
Managerial effectiveness in fisheries and aquaculture
Marine climate, weather and fisheries
Marine fish behaviour in capture and abundance estimation
Marine ecosystems behaviour & management
Marine fisheries ecosystem
Marketing: a practical guide for fish farmers
Marketing in fisheries and aquaculture
Mending of fishing nets
Modern deep sea trawling gear
More Scottish fishing craft and their work
Multilingual dictionary of fish and fish products
Multilingual dictionary of fishing vessels and safety on board
Multilingual dictionary of fishing gear
Multilingual illustrated dictionary of aquatic animals and plants
Multilingual illustrated guide to the world's commercial coldwater fish
Multilingual illustrated guide to the world's commercial warmwater fish
Navigation primer for fishermen
Netting materials for fishing gear
Net work exercises
Operations manual for salmonid farming
Pair trawling and pair seining
Pelagic and semi-pelagic trawling gear
Pelagic fish: the resource and its exploitation
Penaeid shrimps — their biology and management
Planning of aquaculture development
Pollution and freshwater fish
Purse seining manual
Recent advances in aquaculture IV
Recent advances in aquaculture V
Rehabilitation of freshwater fisheries
The rivers handbook, volume 1
The rivers handbook, volume 2
Salmon and trout farming in Norway
Salmon aquaculture
Salmon farming handbook
Salmon in the sea and new enhancement strategies
Scallop and queen fisheries in the British Isles
Scallop farming
Seafood science and technology
Shrimp capture and culture fisheries of the US
Spiny lobster management
Squid jigging from small boats
Stability and trim of fishing vessels and other small ships
The state of the marine environment
Stock assessment in inland fisheries
Stocking and introductions of fish
Sublethal and chronic effects of pollutants on freshwater fish
Textbook of fish culture
Trends in fish utilization
Trends in ichthyology: an international perspective
Trout farming handbook
Wrasse: biology and use in aquaculture

Printed and bound by CPI Group (UK) Ltd, Croydon, CR0 4YY
16/09/2026
14963386-0002